Particle Physics
A Beginner's Guide

ONEWORLD BEGINNER'S GUIDES combine an original, inventive, and engaging approach with expert analysis on subjects ranging from art and history to religion and politics, and everything in between. Innovative and affordable, books in the series are perfect for anyone curious about the way the world works and the big ideas of our time.

Particle Physics
A Beginner's Guide

Brian R Martin

ONEWORLD
OXFORD

A Oneworld Paperback Original

Published by Oneworld Publications 2011

Copyright © Brian R Martin 2011

The moral right of Brian Martin to be identified as the Author
of this work has been asserted by him in accordance with the
Copyright, Designs and Patents Act 1988

ISBN 978-1-85168-786-2

Typeset by Glyph International Ltd., Bangalore, India
Cover design by vaguelymemorable.com
Printed and bound by xxx

Oneworld Publications
185 Banbury Road,
Oxford, OX2 7AR
England

Learn more about Oneworld. Join our mailing list to
find out about our latest titles and special offers at:

www.oneworld-publications.com

Contents

11 Astro-particle physics and cosmology 166

Foreword

Einstein wrote that 'The eternal mystery of the world is its comprehensibility'. The field of particle physics attempts to comprehend the mysteries of the smallest constituents of matter that exist in the world. In this book I give an account of the science, beginning with its emergence from nuclear physics and continuing to current and possible future research.

The fundamental question 'What is matter?' has a very long history. For thousands of years, across many cultures, people have attempted to find an answer. A common theme is the belief that matter is composed of units that cannot be divided into smaller quantities. These are often called *atoms*, a term coined from a Greek word meaning 'something that cannot be divided further', although we now know that atoms are divisible.

Given this lineage, it is unsurprising that the earliest enquiries into the nature of matter were purely philosophical speculations. This quest continues today in particle physics. As in earlier times, it is driven by the twin aims of unification and simplicity: the desire to understand and explain an increasing range of phenomena in terms of a decreasing number of assumptions. Particle physicists focus on deducing answers to basic questions such as 'What are the constituents of matter?' and 'How do particles interact?'

The best current theory of particle physics is the so-called *standard model*, a modest name for the most successful physical theory ever constructed. It resulted from a successful marriage between theoretical invention and ingenious experimentation. In the seventeenth century, the importance of experimental observation came to be appreciated (the motto of the Royal Society, founded in 1660, roughly translates as 'Take nobody's word for it') and in modern science the interplay between theory and experiment is crucial. No theory in science is accepted unless it has overwhelming support from experiment, so in this guide I discuss how particle physics experiments are done, how particles are produced and how they are detected.

The core of this introduction to the field is an account of the rise of the standard model, including how it was constructed and how its assumptions have been tested by experiments. It is remarkable what has been achieved in a relatively short period. I trace how the idea of quarks arose from studies of the particle spectrum, discuss how one of the four forces of nature – the strong interaction – binds quarks together to form the observable particles and the emergence of the theory of these interactions. I complete the story of quarks by describing why it became necessary to postulate the existence of additional quarks and how their existence was eventually demonstrated via another of the forces of nature, the weak interaction. I also discuss how the standard model was completed by uniting the weak interaction with electromagnetism.

When the great German physicist Max Planck entered Munich University in 1875, he was advised not to study science because there was 'nothing left to discover'. History has taught us to be more humble. The standard model is hugely successful in explaining particles and their interactions, but there is still work to be done. Among the important questions still to be answered is how particles acquire mass, although most physicists are confident that this particular problem will soon be solved by new experiments designed to detect the so-called 'Higgs boson', a particle that plays the role of generating masses in the standard model. The standard model also says nothing about gravity, the one force of nature with which we are all familiar. The thrust of research in the twenty-first century has turned to what lies beyond the standard model and how experiments can be conducted to test new theories. To do so will involve studying interactions at energies close to those that existed shortly after the creation of the universe – something not yet fully achievable. Particle physics is becoming entwined with cosmology and questions about the very origins of the universe. As I turn to these areas, I highlight some of the most exciting questions being asked and the most vibrant phenomena being discovered.

For the non-scientist, particle physics entails entering a world of strange, unfamiliar concepts, including quarks, gluons, anti–matter and forces that seem to have little bearing on everyday life. For this reason, the structure of this book does not always follow a strict historical path. Neither does it assign credit to every discovery, as all too often this can be contentious and is best left to the historians. The past century of research in particle physics is full of complexity and twists and turns. I have tried neither to oversimplify the concepts nor to just present the facts of particle physics as they are today. Instead, I have given enough detail so readers can understand how we have arrived at our present state of knowledge and an appreciation of why researchers in the field are excited by current theories and experiments. I have deliberately used the minimum of scientific symbols and only a few simple mathematical expressions; there are no actual calculations in this book. The intention is that mathematical technicalities should not deter anyone with an interest in science from getting an overview of the subject. Finally, there is a detailed glossary where readers may remind themselves of the meaning of any unfamiliar terms.

In preparing this book I have benefited from the helpful advice and support of the editor of the *Beginner's Guide* series, Marsha Fillion and particularly my editor at Oneworld Publications, Robin Dennis. I would also like to thank Tegid Jones, of

University College London and Peter Kalmus, of Queen Mary University of London, for their time in reading and commenting on an earlier draft. Any remaining errors, obscurities, etc., are of course my own.

Brian R. Martin
April 2010

Illustrations

1
Particles and forces

We are all familiar with bulk matter, the stuff that everyday objects, including you and me, are made of and we daily experience the effects on bulk matter of gravity, one of the four fundamental forces of nature. We are aware of the gravitational force because it keeps our feet firmly on the ground and ensures that we wake in the morning to the rising of the Sun. Bulk matter consists of molecules, which are combinations of a relatively small number of chemical elements, where in the modern sense 'element' means a substance that cannot be decomposed by the methods of chemistry.

By far the most common elements in the universe are hydrogen and helium. These are remnants from the creation of the universe in the so-called Big Bang, when the universe rapidly expanded from a very small region of space where matter existed at extremes of density and temperature. Most of the heavier elements we have today were slowly synthesised in stars. Initially this was by the fusing together of lighter elements, starting with hydrogen and helium. Fusion produces the energy of stars, including our own Sun. (The difference between the weight – 'light', 'heavy', etc – of an object and its mass is explained in the insert box on page 2.) The elements are then distributed throughout the universe through the rare spectacular cosmic events called supernovas, where within just a few seconds or so, a star suffers a catastrophic collapse and ejects most of its mass into space. It is a startling fact that the matter that makes up our bodies was once part of distant stars.

MASS AND WEIGHT

Although *mass* and *weight* are often used interchangeably, they are not the same. *Mass* is a measure of 'how much there is of an object' and remains the same wherever the object is in the universe. The mass of an object determines how it reacts when a force is applied to it. *Weight* is the force that acts on a body due to a gravitational attraction. Thus, a body with a given mass weighs less on the surface of the Moon than on the surface of Earth, because the Earth has a far greater mass than the Moon and so generates a far larger gravitational force on the body. On the surface of Earth, weight is very accurately proportional to mass and so we may talk of an object weighing one kilogram, although the kilogram is actually a unit of mass.

You are probably familiar with the periodic table of the chemical elements, the modern form of which is credited to the Russian chemist Dmitri Mendeleev, who developed it in 1869. Mendeleev's original table, which was intended to illustrate recurring – hence 'periodic' – trends in the properties of the elements, has been refined and extended over time as new elements have been discovered. The present table contains ninety-two naturally occurring elements. Some are very common, such as oxygen and carbon, elements essential for life; others are seldom encountered in the everyday world. There are even some extremely rare unstable elements, such as polonium and francium, whose abundances in nature are uncertain because they are too low to have been measured accurately.

Molecules consist of combinations of units of a single unique type for each element, bound together by the second fundamental force of nature, electromagnetism, known to us through the behaviour of electric currents and magnets. Following the Greeks, we still use the word *atom* for the smallest unit of an element.

The atomic model of matter elegantly explained many phenomena observed in the early years of chemistry; by the start of the twentieth century, the masses and dimensions of many atomic elements had been measured. Atoms are almost inconceivably small. As an example, the dot on this letter 'i' has a radius of about 10^{-5} m and contains approximately 10^{11} atoms of carbon. (The notation used for numbers is explained in the inset box below.) To see an individual atom in the dot with the naked eye, the dot would have to be enlarged until it had a diameter of at least 100 m. Similarly for their masses: the heaviest atom is about 5×10^{-25} kg and the lightest less than a hundredth of this. The book you are reading contains at least 10^{26} atoms, an uncountably large number.

NOTATION FOR NUMBERS

Particle physics involves measurements that range from extremely large to extremely small numbers. To avoid writing long strings of zeros, these numbers are written in a compact form as a decimal number between 1 and 9 multiplied by multiples of 10, where 10^n means 1 followed by n zeros and 10^{-n} means 1 divided by the same quantity. For example, 1.2×10^3 means 1200 and 1.2×10^{-3} means 0.0012.

Atoms are divisible

Throughout the nineteenth century, atoms continued to be considered the most fundamental entities that existed. They were believed to be stable and indivisible. But in the final years of that century, two classic results showed this view to be false.

The first was the observation in 1896, by the French physicist Henri Becquerel, that photographic plates he was using were being fogged by radiation from uranium ores, despite the fact that he had wrapped the plates in light-proof paper and placed a metal sheet between the plates and the ores. Becquerel had accidentally discovered *radioactivity*, the phenomenon in which some atoms spontaneously decay and, in so doing, emit other particles. These new particles were fogging his photographic plates. (In physics, 'decay' carries no negative connotations. It simply means that a system is unstable and transforms to another more stable system with a smaller energy.)

There are three distinct types of radioactivity, one of which, beta decay, is caused by the third force of nature, the weak interaction. As observed in beta decay, the intrinsic strength of the weak interaction lies between that of gravity and electromagnetism and is typically one-thousandth of the strength of the latter. Although not directly experienced in our daily lives, the weak interaction controls the rate at which hydrogen is consumed in the Sun and so determines how much heat and light reach the Earth. Ultimately it determines how long life on Earth will be sustainable.

In beta decays, particles called *neutrinos* are emitted. Neutrinos have no electric charge and interact with matter only by the weak interaction and gravity. Because of this, they can pass through many millions of kilometres of material without interacting, making their detection extremely difficult. Their existence was predicted theoretically in 1930, but it took twenty-five years before they were confirmed to exist and a further fifty years to show that they had a mass, albeit a minute one even on the scale of particle masses. We still do not know an accurate value for the neutrino mass, but it is less than 10^{-9} times the mass of the lightest atom.

Some neutrinos are relics of the Big Bang and others are produced in supernovas, but the vast majority of neutrinos detected on Earth come from the reactions that power the Sun. Every second, somewhere between 10^{13} and 10^{14} neutrinos from the Sun pass through our bodies with minimal interactions, doing us no harm. Far smaller numbers are emitted by radioactive rocks in the Earth's crust and a few thousand neutrinos result from the decay of traces of radioactive atoms in our own bodies.

Some unstable atoms decay to other atoms by emitting gamma ray radiation. This is part of the spectrum of electro-magnetic radiation – waves that transmit electromagnetic energy. Electromagnetic radiation was predicted to occur by the Scottish physicist James Clerk Maxwell, who in the nineteenth century constructed a single theory that described both electricity and magnetism. Electromagnetic radiation is most familiar to us as visible light, X-rays, radio waves and microwaves. On the surface, these varieties of radiation appear to be very different. They certainly have distinct physical properties – we see using visible light, whereas exposure to X-rays can destroy our sight. These are a consequence of their different wavelengths and energies, yet each form consists of particles – *photons* – with zero mass and zero electric charge. The discovery that the universe is bathed in an almost uniform background of low-energy microwave radiation – the afterglow of the creative expansion – provided compelling evidence for the Big Bang.

The second blow to the indivisibility of atoms came in 1897 from a study of cathode rays made by the English physicist, J.J. Thomson. Cathode rays can be seen when a powerful battery is connected across two terminals set in a sealed glass tube from which most of the air has been removed. (You may remember that cathode ray technology was used in televisions before the invention of liquid crystal displays, or flat-screen televisions. In a television tube, a cathode emits electrons that are directed on to a glass screen coated with chemicals that produce coloured light when irradiated and creates the picture seen by the viewer.) One of Thomson's experiments with a cathode ray tube showed that the radiation consisted of particles with a negative electric charge: *electrons*. When Thomson measured the mass of the electron, he found it to be unexpectedly small – about two thousand electrons have a mass equal to that of one hydrogen atom. Thomson's discovery of the electron is often said to mark the start of particle physics.

Later experiments established that the electron has two much heavier siblings, the *muon* (or 'mu') with a mass about 200 times that of the electron and the *tauon* (or 'tau') with a mass about 3500 times greater. The electron, muon and tauon, together with the neutrino, constitute a family of particles called *leptons* (from a Greek word meaning 'thin'). Unlike the electron, the muon and tauon are unstable and spontaneously decay via the weak interaction. They play no role in everyday matter and why they appear in nature is a puzzle.

Inside atoms and nuclei

Because stable atoms are by definition electrically neutral, meaning they have zero net electric charge, Thomson's discovery implied that atoms must have an exactly equal positive charge within them. The form of this charge was investigated in a series of experiments carried out in the early part of the twentieth century under the direction of Ernest Rutherford, a brilliant scientist originally from New Zealand, working in England. Rutherford was very wary of elaborate theories and was known to say that he didn't want to hear any physics that couldn't be explained to a barmaid. Contrary to his expectations, Rutherford found that all the positive charge of an atom, and the vast bulk of its mass, were concentrated in a tiny region at the centre of the atom – 'like a gnat in the Albert Hall', as he described it. That gnat is the *nucleus*. Its minute size in relation to the size of the atom can be gauged by the fact that the dot on the 'i' would have to be magnified to at least 5000 km in diameter to be able to see the atom's nucleus with the naked eye.

Rutherford's revelation led the Danish theorist Niels Bohr to construct a 'planetary' model of the atom, in which the nucleus served as the 'sun' and the electrons, akin to 'planets', circled at great distances, not held in their orbits by gravity as in the solar system, but by the electromagnetic interaction. Although this analogy is widely used, it is only qualitatively correct. The ratio of the diameter of the Sun to the diameter of the Earth's orbit is one hundred times that of the diameter of a nucleus to the diameter of an atom – that is, the fraction of 'empty space' in an atom is vastly greater than in the solar system.

Bohr constructed his model using the ideas of the newly-emerging *quantum theory*. One of the things the model explained was why atoms are stable. This was important, because according to both quantum physics and non-quantum physics (or classical physics), charged particles moving in circular orbits continuously radiate electromagnetic energy, thereby losing energy. If this process continued, it would result in the electrons spiralling into smaller and smaller orbits, eventually losing all their energy and leading to the atom's collapse. This would not happen slowly; hydrogen atoms, for example, would have a lifetime of less than one second and without stable hydrogen there would be no universe as we know it. Instead, Bohr postulated that electrons move in fixed orbits around the nucleus and only radiate energy when they shift from one orbit to another. Such transitions successfully explained why electromagnetic radiation from heated atoms appears only at certain distinct wavelengths and not as a continuous spectrum of wavelengths.

Bohr was working at a time when atomic physics was in its infancy; modern quantum theory no longer interprets electrons as orbiting the nucleus in definite circular trajectories with well-defined velocities. (For an introduction to quantum ideas see Alistair Rae's *Quantum Physics: A Beginner's Guide*.) Nevertheless, with appropriate modifications, Bohr's model of a very small central nucleus, with positive charge, surrounded by a cloud of negatively-charged electrons, remains the essential basis of our interpretation of atomic structure, chemistry and biology, so I will continue to use the straightforward language of electron orbits.

The simplest atom is hydrogen. In the Bohr model, it consists of a single electron orbiting a nucleus, which, to ensure the atom as a whole is electrically neutral, must have a single positive electric charge. The nucleus in this case comprises a *proton*, a particle about two thousand times heavier than the electron and with an effective radius of about 10^{-15} m. Heavier elements were initially thought to consist of nuclei with varying numbers of protons orbited by an equal number of electrons, ensuring that atoms were neutral. However, in 1932 the English physicist James Chadwick discovered another constituent of the nucleus. This was an electrically uncharged particle, about 0.1% heavier than the proton, called the *neutron*. This discovery was not unexpected, because Rutherford had already deduced that the nucleus must include uncharged constituents with masses similar to that of the proton and had even coined the name 'neutron'. Protons and neutrons, collectively called *nucleons*, are members of a larger class of particles, *baryons* (from the Greek word meaning 'heavy', because at the time they were the heaviest particles known). We will meet other baryons in later chapters.

The discovery of the neutron was a crucial step in understanding nuclei, including radioactive ones. For example, beta decays are the transformation within a nucleus of a nucleon of one type, either a proton or a neutron, to the other. You may wonder how a proton can decay to a neutron if the neutron is heavier than the proton; conservation of energy would seem to make this impossible. However, while a proton not bound in a nucleus cannot transform to a neutron, it is possible in some circumstances for a proton within a nucleus to do so. This is because the proton can use the additional energy from the force that binds nucleons in the nucleus. Beta decay occurs if it results in the total energy of the final atom, taking into account the energy due to binding, being lower that that of the initial atom. The same applies to a neutron bound in a nucleus, whereas a free neutron can always decay to a proton.

The road to quarks

In the late 1940s and early 50s, more particles were discovered. Some, like the nucleons, were part of the baryon family but had new, unexpected properties. In addition, some unstable, very short-lived particles were found, with lifetimes of about 10^{-23} seconds, far shorter than any particle lifetimes known before.

To understand the behaviour of these very unstable particles, it is helpful to compare them to atoms and nuclei. The state of an atom or nucleus which has the smallest possible energy is called its *ground state*. Atoms and nuclei can be excited into unstable states, called *resonances*, by absorbing external energy, rather like the vibrations of a violin string. When the atom quiets and reverts to its ground state, the energy that was absorbed is released in the form of electromagnetic radiation, for example as visible light. (In the case of the violin, the energy appears as sound.) Excitations occur in nucleons and many other particles; these were the new, very short-lived states that were observed.

The existence of these new particles urgently needed an explanation. It took a decade to understand their complex production and decay patterns, but in 1961 theorists showed that both the ground states and the resonances of the particles could be interpreted as if they were clusters of even smaller particles, named *quarks* by the American physicist Murray Gell-Mann. Quarks are bound together in clusters by the last of the forces of nature, the strong interaction, which is typically forty times stronger than electromagnetism.

At the time, quarks were generally viewed as convenient devices for mathematical modelling and very few physicists believed they were real physical particles. One reason stemmed from the fact that observed particles, such as nucleons, had electric charges whose magnitudes were integer, or whole number, multiples of the charge on the electron, long considered to be the fundamental unit of electric charge. But if nucleons and other particles were composites of quarks, then the model required that quarks had electric charges that were not integer multiples; charges of $\frac{1}{3}$ and $-\frac{2}{3}$ times the magnitude of the electron charge were predicted. Plus, no one had yet been able to detect particles with non-integer electric charges.

These doubts were not dispelled until the late 1960s, when a series of experiments, similar to those conducted by Rutherford, explored the structures within nucleons. The experiments confirmed that protons were indeed composite objects formed from three constituents, whose properties were consistent with those hypothesised for quarks. They also established that if a quark has any size at all it is at least a thousand times smaller than the size of a nucleon, no larger than 10^{-18} m. To return to the dot on the 'i', the dot would need to be magnified to at least ten million kilometres across to have any hope of seeing a single quark in a hadron with the naked eye. However, all attempts to free a quark from a nucleon, or any other particle, have thus far been unsuccessful, unlike in atomic physics, where electrons are easily removed from atoms and nuclear physics, where the nucleus can be split to yield its constituent protons and neutrons.

To understand the structure of nucleons requires just two distinct types of quarks, called 'up', denoted by u and 'down', denoted by d, with the d quark being slightly heavier than the u quark. For example, beta decays are now interpreted as transitions between these quarks bound in nucleons. Later experiments produced a number of new unstable particles that required the existence of a third quark; this was followed by theoretical speculations that a further three quarks should exist, all of which are unstable and all of which, as we shall see, have since been experimentally confirmed. Like the muon and the tauon leptons, the four additional quarks seem to play no role in everyday matter. Presumably these particles were created in the Big Bang but rapidly decayed via the weak interaction and no longer exist naturally in the sense that the electron and the u and d quarks do. Quarks now play a central role in interpreting the phenomena of particle physics.

Forces and fields

We have seen that atoms consist of a tiny nucleus surrounded by even smaller electrons. If this were the whole story, most of the volume of an atom would be 'nothing' and since we are made of atoms, we also mostly would be 'nothing'. So, for that matter, would walls – and nothing would prevent us from walking through a wall! Clearly something is wrong: the answer lies in considering how particles interact.

A satellite circling the Earth is moving in free space but is constrained by the gravitational forces exerted on it by the Earth, the Moon, the Sun and other more distant astronomical bodies. Physicists talk of the intervening space as being 'permeated' by the gravitational fields generated by these bodies. The word *field* is simply a shorthand way of saying that a physical property is assigned to the points of space and time in a region. For example, an interactive weather map showing wind speed at different times of the day represents a field, and the wind speed is the physical property (field strength) assigned to it. We also talk of *potential energy* as the energy stored within the field (because it has the potential to be converted into other forms of energy, such as kinetic energy, the energy associated with motion).

Likewise, the 'free space' in atoms is permeated by the electromagnetic fields generated by the charged particles within the atom. It is these fields that prevent us from walking through walls, though we are mostly 'nothing'. (There are no other significant fields in the atom outside the nucleus. Because the electrons do not experience the strong interaction, there are no strong fields; the forces due to the weak interaction and gravity are far feebler – in the case of gravity by a factor of 10^{40} – and their associated fields are negligible.)

The electromagnetic force is transmitted through the constant exchange of photons between the atom's nucleus and its electrons. Photons serve as 'force carriers' of the electromagnetic interaction, much like a situation where two people 'communicate' by exchanging a ball between one another. If the ball is very light, the people are able to pass it over large distances, but as the mass of the ball increases the distance that can be covered is reduced. This analogy holds for particles: the heavier the particle being exchanged, the shorter the 'range' of the force; that is, the shorter the distance over which the force is significant. In fact, the range of a force is proportional to the inverse of the mass of the exchanged particle. Because photons have zero mass, the electromagnetic force is said to have an infinite range.

The same exchanges occur between each of the protons in the nucleus. Following the rule for electric charges that 'like charges repel and unlike charges attract' would mean that these exchanges between protons would quickly blow the nucleus apart, so there must be an attractive force as a balance. This attractive interaction is the *strong nuclear force*. It is the same for protons and neutrons and is independent of electric charge. Adding neutrons to a nucleus adds mass and increases the attractive binding force without, for the most part, decreasing the stability of the nucleus. The balance between the strong nuclear force and electromagnetism largely determines why the heaviest naturally-occurring stable nucleus is uranium, which has 92 protons and 146 neutrons in its most common form. Particles that experience the strong nuclear interaction, such as nucleons and the short-lived resonances that were discovered in the 1940s and 50s, are known as *hadrons* (from the Greek word for 'thick').

Within hadrons, the weak and strong interactions between quarks also generate fields and have particles that serve as force carriers. The weak interaction has three such particles: two with electrical charge, the W^+ and W^- particles (one has a positive charge and the other a negative charge) and one without a charge, the Z^0. Unlike the photon, these three particles are all very heavy, with masses eighty to ninety times that of a proton. This makes the range of the weak force very short, about 10^{-18} m. The existence of the $W^\pm$ and Z^0 particles was confirmed experimentally in 1983. The strong interaction has eight distinct particles, *gluons*, as its force carriers, all electrically uncharged, so called because they 'glue' quarks together to form hadrons. They are massless and so the strong interaction has an infinite range, similar to the fundamental electromagnetic interaction. The theory of the strong interaction explains why quarks and gluons cannot be observed as free particles; quarks are said to be permanently 'confined' within hadrons.

We seem to have ended up with two strong forces but this is not the case. The strong nuclear force that exists between nucleons, and between other hadrons, is a cumulative effect of the fundamental strong forces between their constituent quarks. It is analogous to the electromagnetic force between atoms, the cumulative effect of the fundamental electromagnetic forces between electrons and protons, the charged constituents in the atom. Just as the electromagnetic force between atoms has a shorter range than the fundamental electromagnetic force, so the strong nuclear force has a shorter range than the fundamental strong interaction between quarks. The strong nuclear force has a range of approximately 10^{-15} m, considerably larger than the range of the weak force.

How can a neutron decay by the weak interaction, when this involves emitting a W particle with a mass eighty times heavier than itself? In the everyday world, it would violate common sense for a loaded truck weighing a total of one tonne to open its doors and jettison an eighty-tonne cargo! In the quantum world, however, common sense is often not a good guide. There is an important constraint in quantum physics, called the *uncertainty principle*, first formulated by the German physicist Werner Heisenberg, which states that energy conservation can be violated but only for a limited period of time. As the energy violation increases, the time period within which 'borrowed' energy has to be 'paid back' decreases. The uncertainty principle is the reason why the range of the weak interaction is so short, because it corresponds to the maximum distance that a W particle can travel before it has to be absorbed by a second particle, in the process 'cancelling' the first energy violation and ensuring that energy is conserved overall.

The various interactions also give rise to characteristic interaction times. For instance, a proton has a radius of about 10^{-15} m. A particle moving at the speed of light (3×10^8 metres per second), would take about 10^{-23} seconds to cross the proton's diameter. If the second particle were also a hadron and thus subject to the strong nuclear interaction, an interaction with the proton should occur within this time. Likewise, if an unstable particle were to decay by the action of the strong nuclear force, its lifetime would be of the order of 10^{-23} seconds, as with the hadron resonances. Similarly, particles with electromagnetic decays typically have lifetimes in the range of 10^{-15} to 10^{-21} seconds and those decaying via the weak interactions have lifetimes in the range of 10^{-7} to 10^{-14}. However, these are only rough estimates; in practice other effects, particularly how much energy is released during the decay, can substantially change actual particle lifetimes. The less energy emitted, the more the decay is suppressed and the more the particle's lifetime is increased. An example is the neutron, which is only 0.1% heavier than the proton. A free neutron decays to a proton plus an electron and a neutrino by the weak interaction, but because the mass difference between the neutron and its decay products is so small, the neutron lifetime is measured in minutes rather than fractions of a second as in other weak interactions.

The past, the present and the future

We have come a long way from the early ideas of atoms as indivisible particles. First, there was the discovery of the electron and then of the nucleus and its constituents, the nucleons. We now know that nucleons are just one example of a class of particles, hadrons, themselves composites of even smaller objects, quarks. Unlike the nucleons within a nucleus, quarks are permanently trapped inside hadrons and cannot be observed as free particles. In addition, there exist two other families of particles: leptons, which have no strong interactions, and the 'force carriers'. This modern view of particles is summarised in Figure 1.1.

The best theory of elementary particles we presently have is the standard model. Curiously this is still called a model, not a theory, even though it is the most successful theory in the history

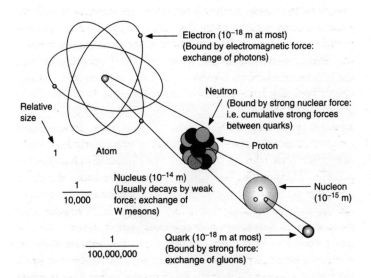

Figure 1.1 The transition from atoms to quarks

of physics. As presently formulated, the standard model is two theories. One operates in the sector of the strong interaction and the other in the sector of the electromagnetic and weak interactions, these two forces having been unified in a way that is equivalent to Maxwell's unification of electricity and magnetism in the nineteenth century.

The standard model aims to explain all the phenomena of particle physics, except those due to gravity (which is so feeble that it plays no role in particle physics at present energies), in terms of the properties and interactions of the three families of particles: leptons, quarks and the force carriers. In addition, at least one other particle, the *Higgs particle*, is postulated to explain the origin of mass. Without this particle, all particles in the theory would be predicted to have zero mass, in obvious contradiction to experimental observations. The particles of the standard model are now referred to as the 'elementary' particles of matter, meaning that they are point-like and without internal structure. A summary of the particles of the standard model and their interactions is shown in Table 1.

In addition to the particles of the standard model, particle physicists also study clusters of quarks – the hadrons. There are

Table 1 Particles of the standard model and their interactions

Particle type	Force experienced	Force mediated
charged leptons	gravity, weak, electromagnetic	
neutral leptons	gravity, weak	
quarks	gravity, weak, electromagnetic, strong	
photon	gravity	electromagnetic
W particles	gravity, weak, electromagnetic	weak
Z particle	gravity, weak	weak
gluons	gravity, strong	strong
Higgs particle	gravity, weak	

several hundreds of these composite states, including nucleons. Hadrons reveal themselves when two particles collide. If the energy of the colliding particles is great enough, then it can be converted into mass that appears as new particles. The equivalence between mass m and energy E is expressed by the iconic equation $E=mc^2$ where the 'exchange rate' is the square of c, the speed of light in a vacuum. Because free quarks are unobservable in nature, physicists are forced to deduce their properties indirectly by studying hadrons, akin to deducing the properties of nucleons exclusively by studying the properties of nuclei. Since nuclei are bound states of nucleons and nucleons are bound states of quarks, the properties of nuclei should, in principle, be deducible from the properties of quarks and their interactions and thus also from the standard model. So far this has proved to be beyond the reach of modern computational techniques: it is similar to saying that we should be able to predict all human behaviour from our knowledge of the body's basic biochemical reactions – in principle possibly true but in practice very distant.

The standard model is hugely impressive in explaining a very wide range of phenomena, but why it has the structure it does is an unanswered question. There are some tantalising clues. One is the possibility that as seen at the energies currently accessible, the three interactions included – the strong, weak and electromagnetic – are manifestations of a single interaction with a single strength. The clear distinctions between them that we observe at present are the remnants of a more symmetrical universe that existed at the ultra-high energies of the Big Bang. At that time, quarks and leptons would not have been immutable and transitions between them would have been possible.

The only certainty is that the standard model is not the final word. It says nothing about why the forces and masses have the values they do, or anything about gravity. There is also strong evidence from cosmology that there are sources of mass and energy that are not included in the model. In the Big Bang scenario, the fate of the universe – whether it will continue to expand, whether the expansion will cease at some time, or whether the universe might even contract in the future – is governed by its average density, because this determines the gravitational forces. Baryons, in all their known forms, make up only a small fraction, about 15%, of the matter in the universe. Scientists are forced to conclude that the bulk of matter, as much as 85%, is non-baryonic. This 'missing' matter is collectively called 'dark matter'. Even more startling, as much as 80% of the energy of the universe (the non-matter contribution), collectively called *dark energy*, has a completely unknown origin. The search for a greater unifying theory of particle physics continues.

A major problem with these extended theories is that they apply at energies well beyond those now accessible. For example, although the particles of the standard model are assumed to be structureless and this is consistent with experimental findings, many physicists believe that if distances could be probed down to the unimaginably small size of 10^{-35} m (yes, thirty-five zeros!), structures would be discovered. Frustratingly, it is difficult to see how this question can ever be studied experimentally, because it would require producing particles with energies comparable to those that existed at the instant of the Big Bang. However, particle physicists can already probe the structure of matter at energies close to those that existed about 10^{-9} seconds after the Big Bang. These experiments will shed light on the very early stages of the universe and enable scientists to construct the story of how the universe was created and evolved. Perhaps then we will finally be able to give a definitive answer to the question, 'What is matter?'

2
Theory and experiment

Discovery is rarely a linear progression towards the truth; neither is the act of discovery always straightforward. The history of science is littered with prior claims to many discoveries, but it is not sufficient simply to have observed a new phenomenon or to have had a new idea. It is also essential that the significance of the discovery is realised and a convincing interpretation given, in a way that leads to further, experimentally-testable, predictions. This interplay of theory and experiment is the essence of the scientific method. In this chapter, I will illustrate it by exploring the relationship between energy and distance and the concepts of *spin* and *anti-particles*. Theories not only have to explain data but they must also be mathematically correct and so I will also give a very brief discussion of quantum field theories, which have played a crucial role in particle physics, and Feynman diagrams, their very useful pictorial representations.

The discovery of the nucleus

Before the discovery of the nucleus, the picture of the atom was that formulated by Thomson following his discovery of the electron. In his 'plum pudding' model, the electrons were embedded in a region of positive charge that filled the whole space occupied by the atom, the electrons being the 'plums' and the diffuse positive charge being the sponge of the 'pudding'.

Rutherford decided to test this model by directly probing the structure of an atom, using particles as projectiles, bouncing, or scattering, them from a target atom. Under his direction, two colleagues, his fellow-New Zealander Ernest Marsden, and the German Hans Geiger, carried out a series of ground-breaking experiments involving radioactive decays. It was already known that in some of these decays, heavy, positively-charged particles, with a mass about four times that of the proton, were emitted. Rutherford named these *alpha particles*, and he used them as the projectiles. The atom chosen was gold, in the form of a very thin foil.

It is worth looking at the apparatus this team used, because it is the prototype of all later scattering experiments in particle physics (see Figure 2.1a). Its main components were a cylindrical metal box (the scattering chamber), about the size of a desktop computer, which contained the source of the alpha particles; the gold–foil target; and a microscope, to which a zinc sulphide screen was rigidly attached. To minimise interactions between the alpha particles and air molecules, most of the air was extracted from within the box. The microscope and the box were fastened to a platform that could be rotated around the foil and the radioactive source, which remained in position. The alpha particles were formed into a beam by passing them through a slit and the beam was directed on to the foil. By rotating the microscope, the distribution of scattered alpha particles – that is, the number deflected by the foil at different angles – could be observed on the screen, which, like a cathode-ray television tube, produced scintillations of light when struck by the scattered particles (see Figure 2.1b). The apparatus was operated by a single person, who had to accustom his eyes to the dark before starting the experiment, so as to be able to spot and count the very weak flashes of light. In modern particle physics experiments, scientists use huge complex instruments the size of buildings that need hundreds, if not thousands, of physicists, engineers, computer

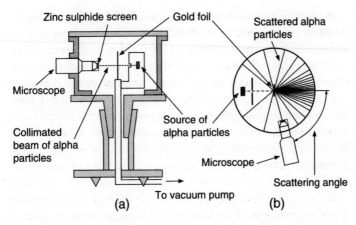

Figure 2.1 Schematic diagram of the apparatus of Geiger and Marsden:

(a) vertical cross-section;

(b) horizontal cross-section of the scattering chamber showing the scattered alpha particles.

scientists and technicians to build and operate them. This inflation is the price to be paid for studying ever smaller and smaller objects.

In the Thomson model, the alpha particles would be deflected a negligible amount by the electrons, because the latter have masses far smaller (by a factor of about 8000) than the alpha particles. This is like bouncing a very heavy ball, such as a gymnasium medicine ball, off a much lighter ball, such as a tennis ball; the direction of the heavy ball would be little changed by the impact. Similarly, the alpha particles were expected to suffer only very small deflections from the positive charge of the gold atoms, since in the Thomson model the charge was distributed throughout the whole atom and effectively diluted.

This is not what Geiger and Marsden observed when they scattered the alpha particles. While small deflections were indeed normal, a small fraction was deflected at very large angles, some almost back in the direction from which they had come. Rutherford described it as 'almost as incredible as if you fired a 15-inch shell at a piece of tissue paper and it came back and hit you'. He interpreted the findings as an indication that the positive charge of an atom and most of its mass are concentrated in a tiny region at the centre of the atom. This discovery opened the way for Bohr's model and the birth of atomic physics.

Rutherford and his colleagues had designed an experiment to test a model, met with unexpected results and used these results to construct a new theoretical model of the atom, with testable predictions that were subsequently confirmed. Why were the alpha particles suitable probes for exploring the structure of an atom? To investigate the structure of an object you need a probe smaller than the object itself. Just as it is impossible to probe the structure of a grain of sand using your finger, it is impossible to study anything smaller than the wavelength of light, typically 10^{-6} m, using an optical microscope. Biological specimens, such as single-celled creatures, can be studied under an optical microscope, but atoms cannot. To test the Thomson model, it was necessary to identify a probe with a wavelength much less than 10^{-10} m, because atoms have dimensions of this size.

Alpha particles satisfied this criterion because of a fundamental property of particles, *wave-particle duality*, that later emerged in quantum theory. All particles behave as waves and conversely, all waves behave as particles. This is true for all objects, even macroscopic ones like you and me, but is only apparent at the quantum level of matter. (In Chapter 1, I anticipated wave-particle duality by referring to electromagnetic waves as photons, i.e. particles.) A striking piece of evidence for wave-particle duality is the observation that a beam of electrons, usually thought of as particles, changes direction when it passes through a crystal. This phenomenon, called *diffraction*, is commonly seen in optics and is normally associated with the wave-like behaviour of light.

In quantum theory there is a very simple relationship between the momentum of a particle, that is, the product of its mass and its velocity, and the wavelength of the corresponding wave. The wavelength is inversely proportional to momentum; doubling the momentum halves the wavelength. Using a probe with a suitably short wavelength is equivalent to using a particle with a suitably large momentum or the equivalent suitably large energy. An approximate guide is $E = 10^{-6}/\lambda$, where the Greek letter lambda (λ) is the wavelength measured in metres and E is the energy measured in electron-volts. (For the units used for energy, see the inset box below.) To explore down to distances of say about 10^{-12} m requires a probe with an energy about one mega-electron-volt. The energies of the alpha particles emitted in radioactive decays fall in this range.

UNITS IN PARTICLE PHYSICS

To make meaningful quantitative comparisons in science we need to ensure that the quantities being compared are expressed in the same units and that these units are an appropriate size for their purpose. Consider for example energy: nutritionists measure energy in calories, while electricity companies use kilowatt-hours. Both of these are related to the joule (J), a unit of energy defined in the International System of Units. In particle physics, the unit used to measure energy is the electron-volt (eV), defined as the energy that an electron carrying one unit of electric charge would have after being accelerated through an electrical potential of one volt. Using $E=mc^2$, we can express particle masses in units of eV/c^2 and particle moments in eV/c, where $1\ eV/c^2 = 1.78 \times 10^{-36}$ kg, a very small mass. Since the eV unit is so small, it is also common to use kilo electron-volts ($1\ keV = 10^3 eV$), mega electron-volts ($1\ MeV = 10^6 eV$), giga electron-volts ($1\ GeV = 10^9 eV$) and tera electron-volts ($1\ TeV = 10^{12} eV$).

Photons and neutrinos

The first suggestion that photons existed emerged in 1900 from work by Max Planck on black-body radiation, the radiation emitted by all bodies when they are heated. The classical description of electromagnetic radiation had spectacularly failed to explain experimental observations: it predicted the nonsensical result that the total power radiated by a heated body was infinite. Planck solved this problem by postulating that a radiating macroscopic body consists of an enormous number of elementary oscillators (now known to be atoms or molecules), each vibrating at some frequency between zero and infinity and that these frequencies could only take certain values, that is, were *quantised*, implying that the energy carried by the emitted electromagnetic radiation was also quantised. Planck claimed he made this radical suggestion almost as an act of desperation, because he had tried everything else without success!

Planck thought the quantisation was a consequence of the emission process, but Albert Einstein quickly demonstrated that it was an intrinsic feature of electromagnetic radiation in general, not just confined to the radiation from heated bodies. For example, Einstein was able to show that quantisation could explain previously puzzling features that occurred when electrons are emitted from the surface of metals bombarded with electro-magnetic radiation. These quantised packets of electromagnetic energy are *photons*. (See the inset box about the photon on page 25.) The gamma rays seen in the decays of some nuclei are photons with a particular range of wavelengths, as are the X-rays emitted when excited atoms decay to their stable ground states. Figure 2.2 shows the frequencies in hertz (Hz), defined as one cycle per second, for a range of wavelengths and the corresponding values of energies.

Figure 2.2 also shows the 'effective' temperature in kelvin. Room temperature is about 293 K (temperature in °C = K −273).

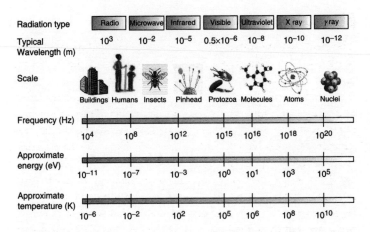

Figure 2.2 The spectrum of electromagnetic radiation

THE PHOTON

Photons, denoted by the Greek letter gamma (γ), are particles with zero mass and zero electric charge. Their energy is given by the formula $E = h\nu$, where $\nu = c/\lambda$ is the frequency of the associated waves predicted by wave-particle duality and h is the so-called Planck constant, which has the tiny numerical value $h = 4 \times 10^{-24}$ GeV s. The quantity $h/2\pi$, where π has the well-known value of approximately 3.14, occurs very frequently in quantum theory and so is assigned a new symbol, $\hbar$.

The effective temperature is the temperature at which radiation of a particular energy (or a particular wavelength) would be most likely to exist – the hotter the radiating body, the greater the energy and the shorter the wavelength of the emitted radiation. Temperature is useful when considering a large collection of photons of a given average energy. For example, about 10^{-9} seconds after the Big Bang, particle energies were in the TeV range, the highest energy that is currently attainable in the laboratory, corresponding to an effective temperature greater than 10^{16} K. About 10^{-6} seconds after the Big Bang the effective temperature had cooled to approximately 10^{13} K, corresponding to energies in the GeV range and nucleons were formed from quarks. Not until about three minutes after the Big Bang had the temperature cooled to around 10^{10} K, corresponding to energies in the MeV region, and helium was formed. It took a further 10^5 years before the temperature fell to a few thousand kelvin, corresponding to energies in the eV region, and neutral atoms were formed.

The popular view of scientists is that they are very open to new ideas, even outlandish ones. While this is true, there is an important proviso: before they are generally accepted the ideas have to be supported by overwhelming experimental evidence. Until this happens, convincing the scientific community can be very difficult. Einstein's hypothesis about photons is a case in point: he had great difficulty in persuading other scientists until in 1923 the American physicist Arthur Compton showed that when light is scattered by a particle it behaves exactly as does a particle of zero mass.

The existence of the neutrino was postulated in 1930 by the Austrian theorist Wolfgang Pauli who was looking for an explanation for the fact that energy did not appear to be conserved in beta decays. (The name, meaning 'little neutron', was coined by the Italian physicist Enrico Fermi.) Before Pauli's hypothesis, beta decay had been viewed as a process whereby an initial 'parent' nucleus decayed to a final 'daughter' nucleus and an electron. If these were the only particles in the final state, they would have to carry off equal and opposite amounts of momentum to conserve momentum as expected. Energy would also have to be conserved and so although the magnitudes of the particle's energies would be different, because the masses of the electron and the daughter nucleus are not the same, they would each have a single unique value. However, experiments showed that the electron had a range of energies, apparently violating energy conservation.

Pauli solved this problem by suggesting that there was a third particle, the neutrino, in the final state. The energy released in beta decays could then be shared among the daughter nucleus, the electron and the neutrino in many different ways, with the electron having an energy ranging from zero to a maximum value that depended on the masses of the two nuclei and that of the neutrino. In practice, the maximum energy of the electron corresponded very closely to the difference in masses of the initial and final nuclei, which implied that the neutrino mass was very small. Once experimental uncertainties were taken into account, the data were actually consistent with the neutrino having a mass of zero or very near zero, just like the photon.

Pauli's hypothesis was initially rejected by most physicists. Bohr was even prepared to consider giving up the cherished principle of conservation of energy rather than embrace the neutrino! It was only in 1956 that the existence of the neutrino was confirmed.

Spin and anti-particles

We have seen that an elementary particle is characterised by its electrical charge, which may be zero. Charge is an example of a *quantum number*. In quantum theory, the full set of quantum numbers defines the state of the particle and, along with its mass, determines its properties. Quantum numbers can only take on a range of discrete values. For example, hadrons have charges that are integer multiples of the charge on the electron; quarks have charges that are multiples of one third of the electron's charge. Such quantum numbers play an important role in particle physics. One is related to angular motion: a child's spinning top has angular momentum due to the distribution of its mass rotating about an axis of rotation. Elementary particles – for example an electron orbiting the nucleus of an atom – can also have this type of 'orbital' angular momentum. In addition, quantum particles possess 'spin angular momentum', also called *spin*. Whereas orbital angular momentum will change depending on the motion of the particle, spin is intrinsic to the particle, present even when the particle is at rest. It was an analysis of the spins of atoms that led Rutherford to deduce the existence of the neutron. Although the word spin is used in everyday speech, as in the example of the spinning top, it invariably means orbital angular momentum in that context.

The concept of spin was originally proposed for electrons in 1924 when Pauli was studying the spectrum of electromagnetic radiation emitted from certain metals when they were excited. At that time he formulated the important 'exclusion principle', which initially stated that no two electrons in a system could simultaneously occupy the same quantum state and thus have the same set of quantum numbers. This principle is at the root of atomic structure. Since additional electrons must occupy new quantum states, each element is distinct and has different chemical properties. In the case of spin (s), the possible values are $s = 0, \frac{1}{2}, 1, \frac{3}{2}, \ldots$, in units of $\hbar$. Spin has a fixed value for particles of a given type; the photon has spin 1, whereas electrons, neutrinos and nucleons all have spin $\frac{1}{2}$.

Particles with half-integer spin are called *fermions* and those with integer spin are called *bosons*, names coined by the renowned English physicist Paul Dirac in 1946. The exclusion principle, although originally formulated for electrons, applies to all fermions but not to bosons. The properties of the two types of particles are fundamentally different. Quantum theory also restricts the values that can result from measurements of spin.

For example, measurements of the component of electron spin along any given direction can yield the values $+\frac{1}{2}$ or $-\frac{1}{2}$ with an equal probability. These are referred to as *spin up* and *spin down*, respectively. Spin values can be added or subtracted provided the sum is never negative. So a system of two particles each with spin $\frac{1}{2}$ can have a total spin of either 0 or 1 and a system of three such particles can have a total spin of $\frac{1}{2}$ or $\frac{3}{2}$.

When Pauli hypothesised the existence of the neutrino, he was motivated by the fact that not only conservation of energy, but also angular momentum, seemed to be violated in beta decay. Without the neutrino, a neutron with spin $\frac{1}{2}$ was apparently decaying to two particles (a proton and an electron), which both had spin $\frac{1}{2}$. Assigning spin $\frac{1}{2}$ to the neutrino resolved this discrepancy.

Spin was originally conceived to explain some experimental data; the origin of spin remained unknown. At first, physicists thought that particles such as the electron could be viewed as literally spinning about their internal axis of rotation, similar to a top, but although this classical physics picture is sometimes useful, it is not consistent with the idea of the electron being point-like. In fact there is no analogy for spin in non-quantum physics. To understand how spin arises we must turn to quantum theory, which at that time was only applicable to particles moving at velocities that were small compared to the speed of light, so-called 'non-relativistic' situations.

Physicists soon tried to extend quantum theory to embrace Einstein's special theory of relativity and include particles moving at any speed. This proved to be difficult. In 1928 Dirac stunningly hypothesised a new equation to describe the behaviour of relativistic electrons, but the equation appeared to have a serious problem when it made predictions about a particle's energy. The total energy of a particle is the combination of its intrinsic energy due to its mass and its kinetic energy. In relativity, these energies combine to produce two values for the total energy E, numerically the same but one positive and the other negative. Yet Dirac's equation for any given momentum turned out to have four solutions. One pair corresponded to an electron with positive energy and the two allowed values of the spin projection $+\frac{1}{2}$ or $-\frac{1}{2}$. The other pair corresponded to an electron with negative total energy and also one of these two spin values. Thus spin emerged naturally as a consequence of combining quantum theory and relativity.

The Dirac equation provides a test for whether a particle with spin $\frac{1}{2}$ really is elementary and without structure. At school, students learn about magnetism using a bar magnet. The product of the strength of its poles times the distance between them is the *magnetic moment* of the magnet. The magnetic field generated by the bar magnet is revealed and mapped in the classroom by sprinkling iron filings around the magnet. Quantum particles can also have a magnetic moment: a charged particle with spin, such as an electron, is equivalent to a tiny electric current and like all electric currents it generates a magnetic field, equivalent to that produced by an intrinsic magnetic moment. Using his equation, Dirac was able to predict the value of the magnetic moment for a point-like spin-$\frac{1}{2}$ particle with electric charge. The measured value for electrons was found to agree with his prediction to one part in ten thousand, providing strong evidence that the electron is without structure. In contrast, the values of the magnetic moments for the nucleons do not obey the Dirac equation, therefore nucleons are not classified as elementary particles. Scientists have since discovered that nucleons are composed of clusters of quarks.

The two solutions of the Dirac equation corresponding to an electron with negative total energy appeared to be nonsensical. Most of the scientific community refused to accept them. There seemed to be nothing that would prevent positive energy electrons in an atom from making transitions to states of ever lower negative energies, which would mean that atoms would be unstable, a catastrophic prediction. Pauli, an acerbic and scathing critic – one of his jibes referred to a physicist as 'So young and already so unknown' – is said to have proposed a second 'Pauli principle': theories should be applied to their authors and Dirac should decay immediately along with his 'absurd' theory! However, Dirac was not discouraged by the apparent unphysical nature of these solutions. 'It is more important to have beauty in one's equations than to have them fit experiment. If there is not complete agreement between the results of one's work and experiment, the discrepancy may well get cleared up with further developments of the theory,' he remarked. In his view the theory was so mathematically beautiful that an explanation would emerge with time – and he was proved right.

The explanation, proposed by Dirac himself, was a simple one: that in nature all the states with negative energy are occupied by electrons. He referred to these as a 'sea' of filled negative energy states. Transitions between, or to, such filled states are forbidden by the exclusion principle and cannot be observed. This was an ingenious solution but it generated a further question: if one of the electrons with negative energy absorbs enough energy to be raised to a state with positive energy, what is the nature of the 'hole' it leaves behind in the 'sea'? The hole would have the characteristics of the absence of both negative electric charge and negative energy and therefore would behave in all respects as a particle with positive electric charge and positive energy. This state is now called the *anti-particle* of the electron; Dirac initially proposed that it should be identified with the proton. This suggestion came under intense, if friendlier, criticism from other eminent scientists, such as Bohr and Fermi. Soon several people, including the American theorist Robert Oppenheimer (who later led the team that built the first nuclear bombs in World War II), showed that the mass of the new state would have to be identical to that of the 'normal' electron and so the prediction followed that a new particle – a positively-charged electron – existed.

Although the conclusion that a positively-charged electron had to exist was inescapable, Dirac was extremely reluctant to say so in his published writings and left it to others, until 1931. The next year, the existence of a positively-charged electron was confirmed by the American physicist Carl Anderson in a classic experiment; he named it the *positron*. Even then Dirac perversely suggested that Anderson's result might be some sort of experimental aberration. Dirac's reluctance may have been an after-effect of an early scientific disappointment. Late in life he said 'Hopes are always accompanied by fears and, in scientific research, the fears are liable to become dominant'. The prediction of anti-particles from the Dirac equation holds true for all fermions but, just as strangely, Dirac had no qualms about explicitly predicting the existence of the anti-proton, a negatively-charged proton that was not discovered experimentally until 1955. The story of spin and anti-particles further illustrates the close relationship between theory and experiment.

Anti-particles are indicated by placing a bar over the symbol for the particle. Thus because p is used for the proton, $\bar{p}$ is used for an anti-proton. There are a few exceptions. For example, the electron and positron are written e^- and e^+, respectively, where the superscript denotes the sign of the particle's electric charge. (The notations used in particle physics are largely historical and not always logical.)

Anti-particles are a general feature of nature and exist for all particles, including bosons, not just the fermions described by the Dirac equation. The modern theory of anti-particles makes no use of Dirac's original idea of a sea of negative energy states and gives equal status to particles and anti-particles, the latter no longer being simply the 'absence' of particles. Dirac was a complex man who had great difficulty communicating. He once said of Bohr: 'We had long talks together, long talks in which Bohr did practically all the talking.' Yet he is recognised as one of the giants of twentieth-century physics. In 1933 he became the youngest theorist ever to receive the Nobel Prize in Physics, a distinction he held for a further twenty-four years.

Quantum field theories and Feynman diagrams

Any successful theory in particle physics must be consistent with the twin foundations of all physics: quantum theory and special relativity. It must also account for the fact that particles interact by radiating other particles, the force carriers of the standard model. Such theories are called *quantum field theories*.

The first reasonably complete quantum field theory in particle physics was formulated by Dirac in 1927 for the electromagnetic interactions of electrons. He posited that charged particles such as electrons interacted by emitting photons that generated an associated electromagnetic field that was 'sensed' by another charged particle. Dirac's calculations assumed that only the simplest possible contributions to a process would be important and that more complicated ones could be neglected. For example, if two electrons interacted, he assumed that the exchange of a single photon would be significantly more important than the exchange of two or more photons. Using this approach, Dirac was able to make predictions for measurable quantities. This method of calculation, *perturbation theory*, whereby more complicated contributions become progressively smaller, is widely used in physics. Its application seemed to be justified in electrodynamics because the strength of the interaction is characterised by a number, without dimensions, called a coupling constant (denoted by the Greek letter alpha α), with a value of approximately $\frac{1}{137}$. The rates for processes where a single photon is exchanged are proportional to α but every additional photon exchange has an additional factor of α ($\frac{1}{137} \times \frac{1}{137}$, etc.) and therefore the contribution of these to the rate should be much smaller than that for the exchange of a single photon.

However, when the corrections were calculated, far from being small, they were all infinite! Clearly there was something wrong. Not until the 1940s did physicists find a solution: the problematic infinities were eliminated by a procedure, *renormalisation*, which involved redefining the charge and mass of the electron that appeared in the theory in terms of their measured values. Technically, this is equivalent to subtracting two infinite terms to produce a small finite remainder, which sounds somewhat dubious but which can be formulated in a mathematically correct way.

The most popular renormalised perturbation method was developed by the American physicist Richard Feynman, who based it on diagrams that make it easy to see the contributions to a given process from the various particle exchanges. Two examples of Feynman diagrams for the interaction of an electron and positron are shown in Figure 2.3. The convention in these diagrams is that fermions are represented by solid lines and photons by wiggly lines. Time runs from left to right. The arrows do not represent the direction of travel of the particle

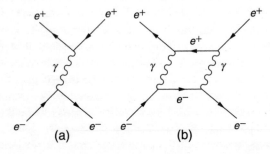

Figure 2.3 Feynman diagrams for the interaction of an electron (e^-) and a positron (e^+) by the exchange of photons (γ)

but instead distinguish particles from anti-particles. Particles have arrows pointing to the right, anti-particles have arrows pointing to the left.

In Figure 2.3a, an electron (e^-) emits a photon (γ) that is then absorbed by a positron (e^+). Thus the electron and positron interact via the exchange of a single photon. This is the simplest contribution. In Figure 2.3b, the electron emits a photon, then a second photon, so that overall the electron and positron interact through the exchange of two photons. The photons in these diagrams are referred to as *virtual* particles because they only exist for times determined by the uncertainty principle and do not appear in laboratory experiments. A factor of α, the coupling constant, is associated with each of the vertices of the diagram, the points where lines meet. Therefore contributions involving multi-photon exchanges are much smaller than those with a one-photon exchange. Feynman diagrams with multi-particle exchanges are referred to as *higher-order diagrams*. Feynman developed the method whereby diagrams like these could be turned into numerical predictions for real data but I will use them only as pictorial illustrations.

The overwhelming majority of particle physicists eventually accepted the renormalisation method because the theory, *quantum electrodynamics* (QED for short), led to predictions that matched experimental data to a very high degree of accuracy. One of these predictions is that a vacuum, conventionally thought of as 'nothing', is actually full of particle-anti-particle pairs that are continually created and annihilated, a seething activity that should have experimental consequences. One such effect is a minute correction to the value of the magnetic moment of a spin-$\frac{1}{2}$ particle, as given by the Dirac equation. Although this was initially dismissed by some, experiments confirmed the correction. Dirac, however, continued to doubt that the renormalisation scheme was the answer, not because it failed to produce results in agreement with experiment, but rather because the technique offended his sense of mathematical beauty. For the rest of his life he remained convinced that there had to be a better, more elegant way of solving the problem of the infinities, although he failed to find one.

Given that the strength of the weak interaction is far smaller than that of the electromagnetic interaction, physicists reasoned that the perturbation approach would lead to a successful field theory of the weak interaction. Fermi had already constructed a theory of the weak interaction in 1935 to analyse nuclear beta decays, using electromagnetism as a guide, but his theory had also been plagued by the problem of infinities. With the discovery of the renormalisation technique, it was hoped that Fermi's theory could be completed. But this turned out not to be the case; the infinities could not be removed. Attempts to construct a quantum field theory of the nuclear strong interaction were also unsuccessful. Interest in quantum field theories was only revived when theories of quarks interacting via the exchange of the force carriers – gluons and the W and Z particles – were formulated.

The three modern quantum field theories of electromagnetism, the weak interaction and the strong interaction, all possess a fundamental symmetry, *gauge invariance*. This means that certain changes can be made to the mathematical quantities that enter the theories without altering their predictions for measurable quantities. For this reason, the force carriers are also called *gauge bosons*. Gauge invariance was first recognised as a property of Maxwell's equations that describe electricity and magnetism in a unified way; it is not just a mathematical curiosity. The transformations are fundamental, because they actually determine the types of interaction that are allowed. In 1971, a seminal piece of research by the Dutch theorist Gerardus 't Hooft proved that gauge theories are renormalisable. This provided the breakthrough that allowed particle physicists to make finite predictions for all measurable quantities and established quantum field theories as the theoretical foundation of the standard model.

There remains the issue of gravity. The infinities encountered in trying to construct a quantum field theory of gravity are far more severe than in the other three forces of nature and there is still no acceptable solution.

3

Accelerators and beams

In a typical experiment, particles are prepared and allowed to interact and the interaction products are observed. The products may be the original particles, scattered in new directions, as in the experiments conducted by Rutherford and his colleagues, a process called elastic scattering. Alternatively, because energy may be converted into matter, the products may be new particles. At high energies this is the most likely outcome, with many particles being produced. Because these new particles are usually unstable and decay rapidly, experimenters study their decay products. In this chapter, I will consider how physicists produce particles to be used as projectiles.

Natural and man-made accelerators

For many years, the only particles available for experiments were those that occurred in nature, with energies that could not be controlled by the experimenter. In most cases the particle's energies were also not unique; only sources with a range of energies were available, such as the products of the radioactive decay of nuclei, as used in the experiments that led to the discovery of the nucleus. Here, the energies are in the MeV region, which is far too low for most modern particle physics experiments, although some experiments are conducted using the enormous flux of neutrinos emitted from nuclear reactors.

These produce energy from the fission of radioactive nuclei, whose decay products are themselves subject to beta decay; the energies of the emitted neutrinos are typically a few MeV.

Much higher energies are available in the secondary particles produced by cosmic rays. Cosmic rays are particles, mainly protons, which impinge on Earth from space; their precise origin is unknown. Some have energies far greater than can be produced in laboratories. They interact with atoms in the atmosphere to produce particles that can be detected at the surface of Earth. Experiments using cosmic rays have yielded many remarkable discoveries, including that of the positron. They were also key to finding the first unstable short-lived particle that does not occur naturally, a boson with spin 0 called the pi meson or more usually just the 'pion' (denoted by the Greek letter pi, π). (*Meson* denotes a boson that experiences the nuclear strong interaction.) Pions exist in three states, one with positive electric charge (π^+), one with negative charge (π^-) and one without charge (π^0).

Although the study of cosmic rays and their products continues to be an interesting subject in its own right, nearly all experiments now use man-made beams produced by accelerators. An accelerator starts with a source of low-energy particles and boosts their energy by applying electric forces. Accelerators can only directly produce beams of stable, electrically-charged particles, such as protons, anti-protons or electrons but there are indirect methods for producing beams of unstable particles and even for producing some neutral particles, including neutrinos.

To begin the acceleration, low-energy particles are injected into an accelerator from a high-intensity source, such as a heated metal filament in the case of electrons. It may typically take several stages, each incrementally boosting the energy, to reach the maximum final energy. Accelerators have great advantages over natural sources of particles: the projectiles produced are of a single type and their energies can be controlled by the experimenter. The earliest accelerators were built in the 1930s but could only produce particles with energies of a few MeV. By the 1950s, energies of few GeV could be reached. The highest-energy machines currently available produce particles with energies of a few TeV and these can explore matter at the effective temperatures that existed very shortly after the Big Bang.

Linear and cyclic accelerators

Early examples of accelerators used static electric fields to accelerate particles. Such machines are limited to relatively low energies and so particle physicists turned to a different technique to produce beams for research: they boosted the initial energy of the particles by repeatedly applying a radio frequency (rf) electric field. A variety of machines use rf fields but the general principles that govern them are similar.

An example of a linear accelerator (also called a linac) using protons is the accelerated particles as shown in Figure 3.1. The protons from a low-energy source pass through a series of metal pipes, called drift tubes, that are located within another pipe, the vacuum pipe, which is maintained at a very high vacuum to minimise interactions between the protons and gas molecules.

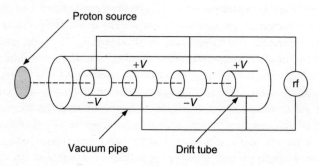

Figure 3.1 Acceleration in a linear proton accelerator

The drift tubes are connected successively to alternate terminals of an oscillator that produces an rf electric field. In Figure 3.1, V denotes the electric potential of the drift tubes and determines the force on the proton. With the signs of the potentials as shown, the protons will be accelerated towards the first drift tube. Because the tube is maintained at a constant potential, the protons will continue to drift through it at a constant velocity. If the rf oscillator can change the field direction (and so the signs of the potentials) as the protons pass through the first tube, the protons will again experience an accelerating force on their way between exiting the first tube and entering the second one, and so on. To be continuously accelerated, the particles must remain synchronised with the oscillating rf field as they pass through each successive tube. Since the speed of the protons increases as they move between the drift tubes, they will take a decreasing amount of time to travel the same fixed distance. If the oscillating field is of constant frequency, each subsequent drift tube must be longer, to ensure that the particles spend the same amount of time in each of the drift tubes. The beam must also remain focused, so that the particles do not hit the sides of the drift tubes during their passage through the accelerator. Proton linacs are often used in particle physics as injectors, that is, they produce proton beams of moderate energy that are injected into a more powerful machine, where they are accelerated to much higher energies.

For electrons, which rapidly approach the speed of light as they are accelerated, a variation of the method is used. The accelerator consists of a straight tube, in the form of a series of cylindrical metal cavities. Power is fed to the accelerator from devices called klystrons, which produce radiation in the form of pulses in the microwave region of the electromagnetic spectrum. These pulses are transported to the accelerator, where they generate an oscillating electric field pointing in the direction of the metal tube (similar to that produced by the rf oscillator in a proton linac) and a magnetic field circling the tube's interior. The metal cavities in the accelerator are of constant length and so the frequency of the microwaves is adjusted to ensure that the electrons arrive at each cavity at the optimal time to receive the maximum energy boost. Vitally, the magnetic field helps to keep the beam focused. The largest linac in the world is the Stanford Linear Collider (SLC) at the Stanford Linear Accelerator Laboratory (SLAC) in California, USA. It consists of eighty thousand copper cavities separated by copper discs, each with a small hole at the centre to direct the beam. The SLC has a maximum energy of 50 GeV. To achieve this energy it is very long – over three kilometres. Remarkably, it is built very close to the infamous San Andreas earthquake fault line!

In contrast to linear accelerators, cyclic accelerators use a circular, or near-circular, configuration. The earliest example is the cyclotron (see Figure 3.2), which consists of two d-shaped sections, contained in a high vacuum, across which an rf electric field is established. The dees are sandwiched horizontally between the poles of a magnet that produces a uniform magnetic field perpendicular to the dees. Charged particles are injected into the

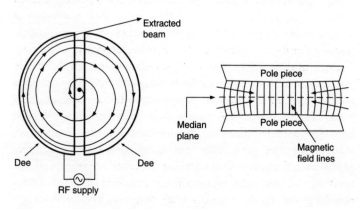

Figure 3.2 Schematic diagram of a cyclotron

machine near its centre, in a plane perpendicular to the magnetic field. Because the particles are moving perpendicular to the field, the force constrains them to follow outward spiral trajectories and the particles are accelerated each time they pass across the gap between the dees. At the maximum radius, which corresponds to the maximum energy, the beam is extracted. The shape of the magnetic field is curved at the edges of the magnet pole pieces. This produces forces, shown as arrows on the diagram, which move particles closer to the median plane, creating a stable beam. The first cyclotron was built by the American physicist Ernest Lawence in 1929 and was a mere thirteen centimetres in diameter.

The cyclotron works on the principle that particles always take the same amount of time to complete a circuit of the accelerator. However, as the speed of the particles increases, the effects of relativity become apparent and that assumption no longer holds; to maintain acceleration, more force is required with each circuit. Over time, the particles cease to be synchronised with the rf field and arrive late at the gap and so do not receive an energy boost. In practice, the cyclotron is only capable of accelerating particles to speeds that are just a few per cent of the speed of light.

Cyclotrons played an important role in the early days of particle physics; Fermi discovered the first hadron resonance using one. Because of their limits, they have since been replaced in particle physics research by cyclic accelerators, *synchrotrons*. Synchrotrons operate similarly to linear accelerators but the acceleration takes place in a circular or near-circular orbit rather than in a straight line. The particles travel in an evacuated tube, the beam pipe, and are kept on their paths by the fields created by an array of magnets that bend the beam, an assembly referred to as the beam line. As the beam repeatedly traverses one or more cavities placed in the circle, where energy is given, the particles are accelerated.

Charged particles travelling in a circular orbit continuously emit electromagnetic radiation, called synchrotron radiation in this context. For a given energy, the losses due to synchrotron radiation increase very rapidly as the mass of the particle decreases and so are very severe for electrons, which are very light. To compensate for the losses, the accelerators input large amounts of rf power, which limits the energies that electron synchrotrons can reach. No similar limit affects electron linacs.

The momentum of an orbiting particle in a synchrotron is proportional to the product of the strength of the magnetic field times the radius of the orbit. Because the momentum increases during acceleration, the magnetic field must be steadily increased if the radius of the orbit is to remain constant, in contrast to the cyclotron, which uses a constant magnetic field. The maximum magnetic field that can be produced over an adequate region is limited and so very-low-temperature technology ('superconducting' or 'cryogenic' technology) is used to reduce power consumption. To achieve very high energies and limit synchrotron radiation, the radius of the accelerator ring must be very large. The Tevatron accelerator at the Fermi National Laboratory in Chicago, which accelerates protons and anti-protons to energies of 1 TeV, has a radius of one kilometre.

In the course of its acceleration, a beam may traverse the ring many millions of times before reaching its maximum energy. Consequently, physicists strive hard to keep the particles in stable orbits; if they meander, they will not achieve optimal acceleration and, if they strike the walls of the vacuum tube, may even be lost from the beam. The particles are accelerated in bunches, each being synchronised with the rf field; oscillations in their trajectories are controlled by another series of magnets, placed at intervals around the beam line, which act like optical lenses. Each magnet focuses the beam in one direction and defocuses it in a direction perpendicular to that, so alternate magnets have their field directions reversed to maintain a stable orbit.

Fixed-target machines and colliders

There are two types of linacs and synchrotrons – 'fixed-target' and 'colliding beam'. In fixed-target machines, particles are accelerated to the highest operating energy and then the beam is extracted from the machine and directed on to a stationary target, usually a solid or a liquid. In this respect, fixed-target machines operate like the earlier cyclotrons. Much higher energies have been achieved for protons than for electrons, because of the large synchrotron radiation losses inherent in electron machines. The intensity of the beam produces large numbers of interactions that can either be studied in their own right or used to produce secondary beams. From $E=mc^2$, it follows that the heavier the particles that are produced, the higher the energy that is required, which reveals a disadvantage of fixed-target experiments when very high energies are needed. When a fast-moving projectile hits a stationary target, some of the energy of the projectile is transferred to the target. This 'energy of motion' of the target is now no longer available to produce new particles and a 'law of diminishing returns' soon sets in, where an increasing fraction of the initial energy is wasted.

This disadvantage may be overcome by using a moving target, to produce a head-on collision between two beams of particles. The SLC and the Tevatron are both colliding beam machines, more commonly known as *colliders*. The two beams often consist of particles and their anti-particles, for example electrons and positrons, circulating in opposite directions in the same machine. The beams are allowed to intersect at a fixed number of points around the accelerator circumference, where experiments are sited. The two beams usually have the same energy and so their combined energy is entirely available for particle production. The products of the collision are scattered rather uniformly without significant momentum in a preferred direction, which has consequences for the way in which particles are detected.

Colliding beam experiments are not without their drawbacks. The colliding particles have to be stable and charged (because they have to be capable of being accelerated by electric fields), which limits the interactions that can be studied. In addition, the probability of a collision in the region where the two beams intersect is far smaller than that achieved in fixed-target experiments, because the particle densities in the beams are very low compared to those in a solid or a liquid target.

Because of the need to produce higher energies in modern particle physics research, almost all the new machines built today are colliders. To date, the largest collider is the Large Hadron Collider (LHC), which collides protons with protons. It has been built deep underground, at depths from 50 to 175 m, in a tunnel at the European Centre for Nuclear Research (CERN), at a cost of approximately £2.8 billion. The LHC is massive, with a circumference of 27 km and straddles the French–Swiss border, near Geneva. Each beam has been designed to have a maximum energy of 7 TeV. Building the tunnel housing the LHC and filling it with equipment was a major international civil engineering project. It became operational in late March 2010, although initially at half the full beam design energy while commissioning tests take place.

Figure 3.3 shows a section of the beam line at the LHC. One of the cylinders is schematically shown open to reveal the two beams of protons, the bending magnets that keep them in circular orbits and the liquid helium system for the magnets. In total, over 1600 superconducting magnets are installed at the LHC, both to bend the beam and to focus it, with most weighing over twenty-seven tonnes. Approximately ninety-six tonnes of liquid helium are needed to keep the magnets at their operating temperature of 1.9 K, making the LHC the largest liquid helium cryogenic facility in the world.

A schematic diagram of the LHC and some of the other accelerators at the CERN site is shown in Figure 3.4. This

Figure 3.3 LHC tunnel and beam line (photo CERN-AC-091118801 by M. Brice; reproduced by permission of CERN)

illustrates the multi-stage process that particle physicists use to obtain high energies. The acceleration begins with a linac, whose beam is boosted in energy in the Proton Synchrotron Booster (PSB) and passed to the Proton Synchrotron (PS), a machine that is also used in lower-energy experiments. The beam is then transported to the Super Proton Synchrotron (SPS), where its energy is increased still further before being injected into the LHC. There are four beam intersection points in the LHC and experiments (denoted by the names ALICE, CMS, LHC-b and ATLAS) are located at each of these points. I will return to some of these experiments and their findings in later chapters. The neutrino beam, marked at the bottom of Figure 3.4, is directed to the Gran Sasso laboratory in Italy, 730 km away. The fact that the beam can travel such large distances is a good example of the very low probability of neutrino interactions mentioned in Chapter 1.

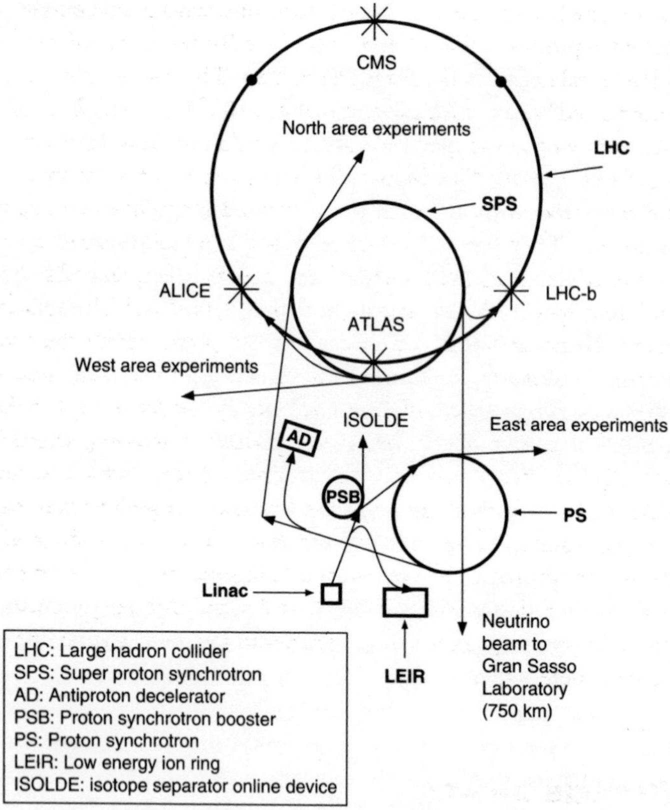

Figure 3.4 Schematic diagram of the accelerators at the CERN site

A particular type of collider has become of increasing interest in recent years. This is the so-called 'particle factory', in which the energies of two colliding beams are tuned so that the total available energy peaks at a particular value corresponding to the mass of a specific particle that physicists wish to study. At present there are two '*B*-factories', one in Japan and one in the United States, that accelerate beams of electrons and positrons, whose total energy is tuned to produce copious numbers of the B-meson, a particle that may reveal details of the weak interaction. The B-meson is a very short-lived boson, with a lifetime of just 10^{-12} seconds. Even if it were moving at the speed of light, it would only travel a distance of about 3×10^{-4} m before decaying, too short a distance to be measured with even the most refined techniques currently available. Therefore, B-factories adopt a novel approach: the beams of electrons and positrons have unequal energies and hence different speeds. As a consequence, the 'debris' (which includes many B-mesons) that is produced at the point where the two beams annihilate is moving in the direction of the faster beam. When quoting particle lifetimes, it is always understood that the particle is at rest. However, if the particle is moving, there is an effect in special relativity called 'time dilation' which means that the lifetime as observed in the laboratory is greater than the 'at rest' value. For this reason, the B-mesons travel sufficiently far in the laboratory to be detected (although the distance is still only measured in millimetres). Essentially, the B-factory sacrifices some energy to produce new particles in favour of the ability to actually detect them.

Particle beams

Accelerators can only directly produce beams of stable charged particles, which in colliders often include anti-particles. Ideally, in a collider, beams of anti-particles should have comparable particle densities to those of the particles with which they collide and also have a single energy. But what is the source of anti-particles? For example, anti-protons could initially be produced in collisions of protons with a target but typically only one anti-proton will be produced for every million protons, so the resulting anti-proton beam would have a very low density and also have a wide spread of momentum, unless a way can be found of 'compressing' the beam.

An ingenious technique for doing this was first suggested by the Dutch applied physicist, Simon van der Meer, in 1968. It uses the electrical signals produced by small groups of particles within a larger bunch to drive an electromagnetic device that applies a correction (a steering pulse or kick) to reduce the momentum spread within the bunch. These individual corrections are applied continuously and over an extended time, so the average tendency of individual particles to move away from the other particles in the beam is reduced and the beam is 'compressed'. This is referred to as 'cooling' the beam. It is appropriate to think of this technique as cooling because the particles in the bunch can be characterised by an internal effective temperature, as we saw when discussing photons. If the average momentum of the bunch were to be subtracted from the momentum of each of its constituent particles, the particles would appear to be moving randomly; the more vigorous the motion, the 'hotter' the bunch, much like the molecules in a gas. Cooling times range from a second to several minutes, depending on the degree of cooling required. Van der Meer applied his technique to anti-protons produced at the SPS at CERN and was able to create an anti-proton beam. This beam was used in a proton-anti-proton collider experiment that discovered the $W^{\pm}$ and Z^0 bosons. Van der Meer shared the 1984 Nobel Prize in Physics for his work on cooling, a rare example of the prize being awarded to an accelerator physicist/engineer.

Physicists are also interested in the interactions of unstable particles, such as pions. Beams of unstable particles can be formed, provided their constituents live long enough to travel appreciable distances in the laboratory. One way of doing this is to direct an extracted primary beam from a synchrotron or linac on to a heavy target. In the resulting interactions with the target nuclei, many new particles are produced. Some of these particles are electrically charged and their trajectories can be guided by applying magnetic fields.

If the new beam is composed of unstable particles, it can be used to produce further beams from their decay products. For example, beams of muons may be made from the decays of pions. There are even ingenious methods to prepare beams of neutrinos from the weak decays of particles such as pions and muons.

4

Detectors
Microscopes for particles

Optical microscopes can only be used to see objects larger than the wavelength of visible light. This limit can be improved by other types of microscope that use electrons instead of light, but even these are incapable of resolving the objects – hadrons and the particles of the standard model – that interest particle physicists. To study these objects, new 'microscopes' have been developed; collectively called *detectors*.

A crucial part of the experiment that discovered the nucleus was a screen coated with zinc sulphide, which produced scintillations of light when struck by scattered alpha particles. Another detector, with which you are doubtless familiar, is the Geiger counter, in which the effect of radiation from a radioactive source is converted into an audible signal, the rate of 'clicks' indicating the strength of the source. This hand-held device is now mostly used to detect possible radiation leaks. Some detectors still use the basic principles behind these two instruments, while others use entirely different mechanisms.

Building a detector

To be detected, a particle must undergo an interaction with the material of a detector. Particles interact with matter in a variety of ways; which is the most important depends on the type of particle and its energy.

The first possibility is that the particle interacts with an atomic nucleus. If it is a hadron, it will do so via the short-range nuclear strong interaction. Because of the charge independence of the nuclear strong interaction, interactions with nuclei are as important for neutral particles as for charged ones. If the energy is sufficiently high, many reactions are possible, most of them involving the production of several particles. Neutrinos and anti-neutrinos can also interact with nuclei but such processes are due to the weak interactions and so the probability of an interaction is extremely small.

In addition to these short-range interactions, when charged particles go through matter, the electric field produced by the nuclei will act on the charge on the particles and accelerate and decelerate them as they pass, causing them to radiate photons and lose energy. This process is called *bremsstrahlung* (German: 'braking radiation') and is a particularly important contribution to the energy loss for electrons and positrons.

A charged particle may also lose energy by exciting atoms along its path or even removing one or more electrons from the atom, leaving a positively-charged atom; an *ion*. These are called *ionisation energy losses* and are proportional to the square of the charge of the particle. A fractionally-charged particle would therefore have a much lower rate of energy loss than any particle with an integer charge. Ionisation and radiation energy losses result from the long-range electromagnetic interaction. They form the basis of most detectors for charged particles.

Radiation losses completely dominate the energy losses for electrons and positrons at high energies but are much smaller than ionisation losses for all other particles at lower energies. Radiation losses decrease as the mass of the radiating particle increases. For this reason, muons penetrate vastly greater thicknesses of matter than electrons of the same energy. Finally, there are photons. In contrast to heavy charged particles, photons have a high probability of being absorbed or scattered through large angles by the atoms in matter. There are several processes that contribute to the overall energy loss but at high energies the dominant effect is the production of electron-positron pairs.

The detection of a particle means more than simply knowing where it is. To be useful, this localisation must be made with a precision, or resolution, good enough to enable particles to be separated in both space and time, to determine which are associated with a particular interaction. Physicists need to be able to identify each particle type and measure its energy and momentum. No single detector is the best for all these require-ments, although some are multi-functional. Modern particle physics experiments commonly use very large multi-component detectors that integrate many different sub-detectors in a single device to make detailed measurements of complex interactions involving even very short-lived particles. A brief overview of detectors is given below, each of which is discussed more detail in later sections.

In early detectors, the passage of a charged particle was converted to a visual track in some way. Many important discoveries were made using these devices but they have now been replaced by electronic detectors. For example, there is a large family of gas detectors that convert the ionisation produced by the passage of a charged particle through a gas into an electrical signal. Such detectors are primarily used to provide accurate measurements of a particle's position or, by a sequence of such measurements, a record of its trajectory. In the latter context, they are also called *track chambers*. Track chambers are very often placed in a magnetic field, in which case the trajectories of charged particles will curve under the action of the magnetic force. The electric charge on the particle is deduced from the direction of curvature, if the direction the particle is travelling and the direction of the magnetic field are both known. Measurements of how much the particle has curved provide a measurement of the particle's momentum.

Spectrometers are dedicated to measuring momentum. They consist of a magnet surrounding a series of detectors to track the passage of the particles. In colliding beam experiments, the reaction products emerge at all angles; the spectrometers must completely surround the interaction region to obtain full coverage. Collection of ionisation products is also possible in a type of solid, a *semiconductor*, whose electrical properties are mid-way between that of a conductor and an insulator. This detector is the solid-state equivalent of a gas detector.

Two other types of detectors are important for charged particles: *scintillation counters* and *Čerenkov detectors*. The zinc sulphide screen used in the discovery of the nucleus is an example of a scintillation counter. Scintillators have excellent time resolution and are often used to decide whether to activate other detectors or whether to record the information from a particular interaction. Without this 'triggering', an experiment to look for a particular type of interaction could be overwhelmed with unwanted data, but making the selection too precise risks missing unexpected phenomena. Čerenkov detectors measure the velocity of a charged particle and can distinguish between different particles with a common, very high momentum by using their velocities to determine their masses from their momenta.

These devices can only detect charged particles. In contrast, *calorimeters* can detect both charged and neutral particles by totally absorbing the particle to yield a measurement of its energy.

Emulsions, cloud chambers and bubble chambers

The oldest visual detector is photographic *emulsion*, a thick layer of a gelatinous material spread on a glass plate. The ionisation energy of a particle passing through the emulsion causes a latent image to form. When the plate is developed, silver grains in the emulsion emerge to show a visual record of the particle's path, in which the thickness of the track increases as the particle's charge increases. While this can be observed and tracked under a microscope, the process is laboriously slow. Emulsions are now largely of historical interest, although are they still occasionally used, in conjunction with electronic detectors, for very special purposes.

One of the most important discoveries using emulsions was made by a group at Bristol University. In 1947, this group was the first to detect charged pions in a cosmic ray experiment. Their discovery was not unexpected; the Japanese theorist, Hideki Yukawa, had predicted their existence in 1935. Yukawa had worked out the theory of the relationship between the range of a force and the mass of the exchanged particle in the context of the strong nuclear force. Using the range of this force, which was approximately known from nuclear physics experiments, he had predicted that the exchange particle responsible for the force would have a mass of about 140 MeV/c^2. This was the first attempt to relate forces to the exchange of particles.

There was therefore considerable excitement in 1937 when a candidate particle was found in experiments using cosmic rays. The particle's mass was somewhat smaller that that predicted by Yukawa but this discrepancy was not considered too serious, because the range of the strong nuclear force was not that well known at the time. However, any particle that is the force carrier of the nuclear strong interaction must, by definition, interact strongly with nucleons; the experiments showed that the new particle could penetrate considerable thicknesses of matter without significant interactions. It took ten years for the Yukawa particle with the predicted mass and properties to be discovered. The lighter particle was the muon, one of the heavier siblings of the electron. It has a mass of 106 MeV/c^2, two hundred times the mass of the electron. Unlike the later discovery of the pion, the discovery of the muon was completely unexpected.

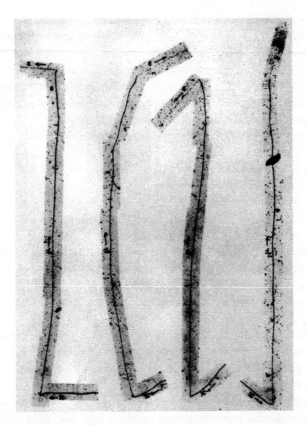

Figure 4.1 Early examples of positively charged pions seen in emulsions

Some examples of pion decay as seen in an emulsion are shown in Figure 4.1. The pions at the bottom of the picture come to rest before decaying to a muon and a neutrino. This is written in the very useful shorthand notation $\pi^+ \rightarrow \mu^+ + \nu$, where Greek letters are used for the particles (π for pion, μ for muon and ν for neutrino.) The muons all have the same energy because there are only two particles in the final products of the decay. Thus the muons each travel approximately the same distance (about six hundred millimetres) in the emulsion before coming to rest and then decaying, this time to three particles – a positron, a neutrino and an anti-neutrino ($\mu^+ \rightarrow e^+ + \nu + \bar{\nu}$). Only the charged particles could be observed and that is why, at the points where particles decay, the tracks have a kink in them.

Another detector, almost as old as the emulsion, is the *cloud chamber*. This detector's operation is based on the observation that the condensation of water vapour into droplets happens more quickly in the presence of ions. A cloud chamber is a vessel fitted with an expansion piston and filled with air almost saturated (holding almost as much as it possibly can) with water vapour. As the user rapidly withdraws the piston, the air in the vessel expands and cools so that it becomes 'super-saturated', that is, has more water vapour in it than it could normally hold at that pressure; an unstable condition. Ions produced by charged particles passing through the chamber can then seed the formation of droplets, which appear along the trails of the particle. The chamber is illuminated by a flash of light immediately after the expansion; if a particle has passed through the chamber, the tracks of the droplets revealed are photographed (through a window in the chamber) before they disappear.

Expanding and recompressing the air with a mechanical piston is very slow; taking about a minute. Since the accelerators used in conjunction with cloud chambers produced bunches of particles every few seconds, it was very inefficient to expand the chamber if no interaction had occurred. To overcome this, it was common practice to place Geiger counters above and below the chamber and only expand the air if the counters registered that a particle had passed through them. This was a key difference from the emulsion, which is continuously sensitive.

Figure 4.2 shows a cloud chamber photograph of one of the first positron tracks observed by Anderson. The band across the

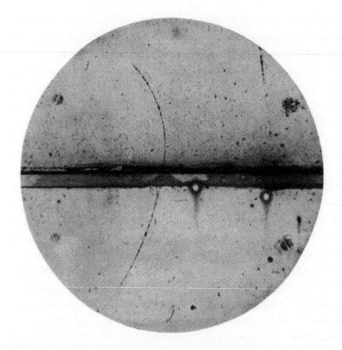

Figure 4.2 One of the first positron tracks as observed in a cloud chamber

centre of the picture is a six millimetre-thick lead plate, inserted to slow down the particles. The track is curved due to an applied magnetic field. The degree of curvature increases as momentum decreases, because the path of a slow-moving particle is easier to bend than that of a fast-moving one. In this case, we can conclude that the particle entered at the bottom of the picture and travelled upwards. The sign of the particle's charge then follows from the direction of the curvature; it is positive.

The track might have been created by a proton but Anderson ruled out this possibility by measuring the range of the upper track. The rate of energy loss of a charged particle in matter depends on its charge and velocity. From the curvature of the tracks, he deduced that the momentum of the upper track is 23 MeV/c, corresponding either to a slow-moving proton or a rapidly-moving very light particle. The proton would lose energy rapidly, coming to rest in a distance of about five millimetres, very close to the thickness of the lead plate. However, the track Anderson observed was more than five centimetres long, enabling him to set a limit on the mass of the particle, that turned out to be compatible with the mass of the electron. This established the discovery of a positively-charged electron, which Anderson named the *positron*. (As it happens, particles that curved in the 'wrong' direction had been seen in the 1920s but had not been taken seriously.) Although Anderson seems not to have been aware of Dirac's prediction of anti-particles, others quickly made the connection.

Because a cloud chamber is filled with air, the rate of collisions is very low. However, by the late 1950s, cloud chambers had largely been superseded by a similar device, the *bubble chamber*. Bubble chambers use a liquid as the working medium, which gives higher collision rates. In a bubble chamber, the liquid is initially held in a vessel at a pressure above the equilibrium vapour pressure at that temperature; that is, above the pressure at which the rate of evaporation of the liquid is the same as the rate of condensation of the vapour. When the pressure is suddenly reduced, the liquid is left in a 'superheated' state (comparable to the supersaturated air in the cloud chamber). The temperature of the superheated liquid is above the boiling point at the reduced pressure, an unstable condition. The passage of a charged particle through the liquid produces ion pairs that act as seeds, with bubbles tending to form along the ionisation trails.

In the heyday of the bubble chamber, the liquid used was often hydrogen. However, as this was a little like working next to a potential bomb, great care was required to prevent explosions! Some experiments used heavy liquids, such as propane, to increase the rate of interactions. Just as for the cloud chamber, it was usual to place the whole bubble chamber in a magnetic field so that the particles' momenta could be obtained; triggering was employed to select particle events of interest.

Gas and semiconductor detectors

Gas detectors record the ionisation produced by the passage of a charged particle through a gas (typically an unreactive gas, such as argon). They collect the ionisation products on to electrodes (or in early experiments made the ionisation tracks visible in some form). In practice, the output is an electrical pulse, which is then amplified.

In understanding the principles of gas detectors, the crucial quantity is the number of ion pairs produced per incoming charged particle for a given voltage applied between the electrodes. At low applied voltages, the output signal is very small, because the electron-ion pairs recombine before reaching the electrodes. As the voltage increases, the number of pairs increases; eventually the electric field strength becomes strong enough for the electron-ion pairs from the primary ionisation to gain sufficient energy to themselves cause further ionisation. This leads to a rapid increase in amplification. The output signal at the positive electrode is proportional to the energy lost by the original particle.

The earliest gas detectors were cylindrical tubes filled with gas and maintained at a negative potential (the cathode) and a single fine central wire at a positive potential (the anode). A major breakthrough was made in 1968, with the discovery that the spatial precision of the detector could be greatly improved by arranging many anode wires in a plane between a common pair of cathode plates, each wire acting as an independent detector. This device is called a *multi-wire proportional chamber*. Even better resolutions are obtained in a *drift chamber*, which takes advantage of the fact that the liberated electrons take time to drift from their point of production to the anode. The time delay between the passage of a charged particle through the chamber and the creation of a pulse at the anode is related to the distance between the particle's trajectory and the anode wire; measurements of time are used to provide measurements of distance. Because time can be very accurately measured, accurate spatial and temporal resolutions can be achieved.

Figure 4.3 shows an example of an event recorded in a cylindrical drift chamber at the Collider Detector at Fermilab (CDF). The detector surrounds the interaction region and the view is along the beam pipe. The diagram shows the electronic reconstruction of the tracks of particles produced by the annihilation of protons and anti-protons. The tracks are curved due to the presence of an external magnetic field.

There is also a solid-state version of the gas detector that uses a semiconductor, usually silicon. In this, the passage of a charged particle raises an electron from a region where it is bound and so immobile, to a state in a region in which it is mobile,

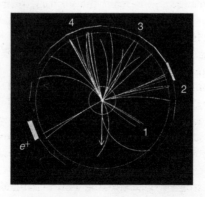

Figure 4.3 Reconstruction of an event in the CDF drift chamber

leaving behind a 'hole'. The resulting electron–hole pairs play the role of electron–ion pairs in gas detectors. In the presence of an electric field, the electrons and holes separate and collect at the electrodes, giving a signal proportional to the energy loss of the incident–charged particle. The energy needed to create this pair is only about a tenth of that required in a gas detector, so a very large number of electron–hole pairs are produced by a low-energy particle, making solid-state detectors well suited for detecting these particles. When they are placed close to the interaction point in a colliding beam experiment to study events involving the decay of very short-lived particles, they are called *vertex detectors*. They are also essential in studying the properties of states containing the heavy quarks, whose lifetimes can be as short as 10^{-13} seconds.

Scintillation counters, Čerenkov counters and calorimeters

In scintillators, the energy loss that occurs from the excitation of atomic electrons in the medium of the detector re-emerges during de-excitation as visible light (or sometimes in the ultra-violet region of the electromagnetic spectrum). In a scintillation counter, this light is directed on to the face of a *photodetector* – a device that converts a weak photon signal to a detectable electrical impulse. An important example of a photodetector is the photomultiplier tube, in which electrons are directed on to the photomultiplier's cathode, from which electrons are ejected. The newly-freed electrons strike a series of devices that amplify and accelerate them. The final signal is extracted from the anode at the end of the tube. The scintillation counter is used widely for triggering other detectors and can be used to provide a starting reference time in a drift chamber. Commonly used scintillators are inorganic single crystals or organic liquids and plastics. Some modern detectors use several tonnes of detector in combination with thousands of photomultiplier tubes.

Another detector, the Čerenkov counter, measures the velocity of a high-energy particle. This counter is based on an effect similar to the sonic boom heard when an aeroplane exceeds the speed of sound in air, although in a Čerenkov counter it is the speed of light that is the relevant quantity. It is important to remember that the speed of light depends on the medium in which the light is travelling. Although it is not possible for a particle to exceed the speed of light in a vacuum, it is possible for its speed to exceed the speed of light in the more normal physical world. With a particle, part of the excitation energy reappears as a cone of electromagnetic radiation. This is the *Čerenkov effect*. The radiation is emitted at a particular angle in relation to the direction of motion, depending on the particle's speed. An image of the ring of the cone is constructed electronically and by measuring the angle of emission, the particle's velocity can be ascertained. The main limitation is that very few photons are produced per centimetre, typically only about one hundredth of the number produced by a typical scintillator.

Calorimeters are important for measuring the energy and position of a particle by its total absorption. A calorimeter differs from most other types of detector in that the detector actually changes the nature of the particle. Some calorimeters are built like sandwiches, with separate layers of dense material (for example lead) that slow down and absorb the particle to be detected and another material (such as a scintillator) that records their energy. Other calorimeters combine these functions in a single material. During absorption, a particle interacts with the material of the detector, generating secondary particles, which will themselves generate further particles and so on, culminating in a shower of particles, just as in a gas detector. Eventually all, or almost all, the primary energy is deposited in the calorimeter, which gives a signal in the detector part of the device. For example, an electron loses energy by radiating photons in a layer of absorber. Then the photons themselves lose energy in the absorber by producing pairs of electrons and positrons. These secondary electrons and positrons then radiate more photons and this continues until the energies of the secondary electrons and positrons fall below the critical energy where ionisation losses equal those from the radiation of photons. At each stage the particles pass through a detector layer so that their energy can be measured.

It is possible to design calorimeters that preferentially detect just one class of particle (say, electrons and photons or hadrons) and it is common to have both types in one experiment, with a hadron calorimeter stacked behind an electromagnetic one. Hadronic showers are qualitatively similar to electromagnetic ones but develop with far more complexity. Many different processes contribute to the production of secondary hadrons but not all the contributions to the total absorption give rise to an observable signal in the detector, as in nuclear excitation or leakage of secondary muons and neutrinos from the calorimeter. Overall, energy measurements for hadrons are much poorer than for electrons and photons.

Multi-component detectors

Modern experiments in particle physics use very large multi-component detectors that integrate many different sub-detectors. One such is the ATLAS detector at the LHC proton-proton collider, at CERN. (ATLAS stands for *A T*oroidal *LHC Apparatu*S – particle physicists, like the military, love acronyms, however contrived!) It is enormous: at about 25 m in diameter and 46 m long, it would occupy about half the space of the Cathedral of Notre Dame in Paris. It weighs about the same as a hundred (empty) Boeing-747 jets. Its scale can be gauged by the human figures shown at the bottom of Figure 4.4. ATLAS's size ensures that the numerous particles produced in the collisions are retained within the detector.

Like all detector systems used in colliders, the sub-detectors are arranged in concentric layers surrounding the beam pipe, because the particles resulting from the collisions of the two beams emerge in all directions. The inner detector consists of a silicon vertex detector, placed very close to the region where the two beams collide, to detect very short-lived particles, plus a number of tracking detectors that can measure particle tracks to a precision of 0.01 mm. These are contained within a magnetic field, to measure the momenta of charged particles from the curvature of their tracks. An electromagnetic and then a hadronic calorimeter are placed outside this field, to measure the energies of particles by absorption. Very large muon detectors are positioned at the outermost sections of the apparatus, because muons are the most penetrating charged particles produced.

The photograph in Figure 4.4. shows a view along the beam direction during the construction of ATLAS. The eight barrel toroids that produce the magnetic field are shown installed, with a calorimeter positioned at the end before its move into the middle of the detector. In the completed detector, the central cavern, where the man is standing, is filled with the sub-detectors. If you compare the ATLAS detector with the apparatus of Geiger and Marsden shown in Figure 2.1, you will see how experiments have developed enormously in the past one hundred years.

Like all modern detector systems, ATLAS relies heavily on fast electronics and computers to monitor and control the sub-detectors and to co-ordinate, classify and record the vast amounts of information flowing in from different parts of the apparatus. If all the data were recorded, this would be equivalent to writing one hundred thousand CDs per second! In practice,

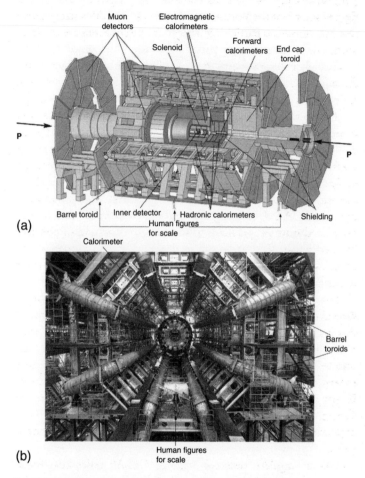

Figure 4.4 ATLAS detector at the *pp* collider LHC at CERN:

(a) schematic diagram of completed detector;

(b) view during construction

the information recorded is equivalent to about thirty CDs per second. The ATLAS detector has many goals, including finding the Higgs boson (if it exists) and so helping to solve one of the outstanding problems in particle physics – the origin of mass. ATLAS is expected to provide data for the next ten to fifteen years.

5

Hadrons and the quark model

After the discovery of new hadron states in the 1950s and 60s, particle physicists attempted to describe all hadrons as composites of other states. This eventually led to the development of the *quark model* of particles. This model raised a number of big questions, the most important of which was 'Where are the quarks?' After a decade of investigation, that question was answered.

Strange particles

Shortly after the pion was discovered, other particles were found in cosmic ray experiments that, like the muon, had not been predicted by theory. A group at Manchester University was the first to make these discoveries; they referred to the new particles as V particles because of the tracks they produced in the detectors.

Two early examples of these events are shown in Figure 5.1. On the left, a neutral particle decays to two charged particles in a cloud chamber; on the right, a charged particle enters the chamber and decays to another charged particle that easily passes through the absorbing plate in the centre of the chamber. This suggests it could be a muon. The kink in the track at the decay point shows that at least one other (neutral) particle has been emitted but it has not been detected directly, because it does not produce ionisation. This is a neutrino. Later experiments using emulsions found evidence for the decay of these V particles to three pions in the

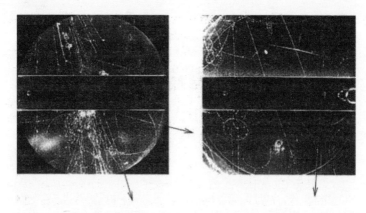

Figure 5.1 Examples of the tracks of V particles as seen in a cloud chamber (Rochester, G.D. and Butler, C.C. 1947. *Nature*, **160**, 855; reproduced by permission from *Nature*, Macmillan Publishers Ltd., copyright 1947)

final state. These experiments established the existence of a new particle in three charged forms (positive, negative and neutral) with a mass that was estimated to be approximately 500 MeV/c^2. We now call these *K-mesons* or *kaons*, with the states denoted as K^+, K^- and K^o. K^+ and K^- are anti-particles of each other. The neutral kaon also has an anti-particle, denoted $\bar{K}^o$, although at this point in our story there is nothing to distinguish it from the K^o.

In the early 1950s, further particles were discovered, all heavier than a nucleon, now collectively referred to as *hyperons*. Hyperons, like nucleons, are part of the baryon group of particles. Like the nucleons, they experience the nuclear strong interaction and so are also members of the larger family of hadrons. There are three distinct types of hyperon, denoted by Greek letters: lambda particles (Λ), which only have a neutral form; sigma particles (Σ), which exist in three charged forms (Σ^+, Σ^- and Σ^0) and xi particles (Ξ), which exist in two charged forms, Ξ^- and Ξ^0. Each of the hyperons also has an anti-particle.

All the hyperons are unstable and decay to a nucleon (or in the case of the Σ^0 to a Λ), plus one or more other particles. Decays that do not result in another baryon in the final state are not observed. To particle physicists this is a clear indication that a law of nature is forbidding these decays. The operation of such a law was already apparent from the observed decays of nucleons. For example, the proton does not decay to a positron and a pion, although there appears to be nothing to forbid it. To codify these observations, baryons are assigned a quantum number, the *baryon number*. This number is an example of an *internal* quantum number, as it does not depend on the space-time properties of the particle. Electric charge is also an internal quantum number.

Baryon number is conserved in all interactions; the total baryon number (obtained by adding the individual baryon numbers of the particles) before the interaction is the same as the total baryon number after the interaction, just as in the conservation of electric charge. In practice, the baryon number is defined to be $+1$ for all the baryons, -1 for their anti-particles and 0 for all other particles. For example, in nuclear beta decay, where a nucleon of one type changes to a nucleon of another type, the total number of baryons is conserved. Likewise, the most likely decay of the lambda particle is to a proton and a negatively charged pion (that is, $\Lambda \rightarrow p + \pi^-$) and baryon number is also unchanged. After the anti-proton was discovered, it was observed that when a proton and an anti-proton annihilate, it is always to a system of particles whose total baryon number is 0 (that is, $p + \bar{p} \rightarrow \pi^+ + \pi^- + \pi^0$). Baryon number differentiates the neutron from the anti-neutron.

One of the most intriguing properties of these new states concerned their decay properties. Although they were produced by the strong interaction at about the same rate at which pions are produced, their lifetimes were far longer than would be expected if they also decayed via this interaction. For example, the lifetime of the charged kaons is about 10^{-8} seconds, characteristic of a weak interaction. Charged pions decay via the weak interaction because there are no lighter mesons to which they could decay via the strong interaction. However, there was no obvious reason why kaons should not decay to two pions via the strong interactions, which would give them a much shorter lifetime of 10^{-23} seconds.

When it came to baryon number, it appeared that a new rule was operating. Gell-Mann coined the name 'strangeness' (S) for this new quantum number because of these particles' unexpected behaviour. The assigned values of the strangeness quantum numbers are $S = +1$ for the K^+ and K^0, $S = -1$ for the Λ and Σ hyperons and $S = -2$ for the Ξ hyperons, with their anti-particles having the opposite sign of S. All other particles are assigned the value $S = 0$. (This choice of values for the strangeness quantum numbers is an historical accident. With hindsight, using numbers with the opposite signs would have been more logical.) The strangeness quantum number distinguishes the K^0 from its anti-particle, the $\overline{K}^0$, and leads to the differences in their properties.

Once strangeness numbers were assigned, selection rules were postulated: strangeness is conserved in strong and electromagnetic interactions but in weak interactions it could change, though by at most one unit. When S does change in the weak interaction, particle physicists talk of strangeness being violated. The rules and assignments attached to strangeness are consistent with all the observed decays: for example, the negative Ξ decays to a Λ particle and a negative pion ($\Xi^- \rightarrow \Lambda + \pi^-$), with a change of strangeness of one unit and not to a neutron and a pion ($\Xi^- \not\rightarrow n + \pi^-$), which would require a change of strangeness of two units. The rules also explained another observation: strange particles, as they are called, are only produced in pairs in the strong interactions of pions and nucleons, one with positive strangeness and the other with negative strangeness, a phenomenon called *associated production*. Thus, positive kaons are produced in association with sigma hyperons ($\pi^+ + p \rightarrow K^+ + \Sigma^+$) but not with protons ($\pi^+ + p \nrightarrow K^+ + p$).

While these rules for strangeness seem almost obvious to particle physicists today, it took five years to deduce them from the various production and decay data. Important steps included the prediction, made by Gell Mann in 1955, of the existence of the neutral states of the sigma and xi (Σ^0 and Ξ^0) from the known properties of their observed charged partners and their experimental discovery four years later. Incidentally, because there is a lighter neutral hyperon with $S = -1$, the lambda, the neutral sigma decays preferentially to a lambda particle and a photon by the electromagnetic interaction, rather than to a nucleon by the weak interaction. For all the other hyperons, there are no lighter states with the same strangeness quantum number and so they are forced to decay via the weak interaction. The lifetime of the neutral sigma is therefore much shorter than that of other hyperons.

Resonances

The hadrons I have discussed so far are analogous to the ground states of atoms or nuclei. Just as they can have excited states, excited states of hadrons also exist. These are unstable and in general decay via the strong interaction to their ground states, with characteristic lifetimes, unless something prevents this from occurring. The first resonance was discovered by Fermi and his group, working in Chicago in the early 1950s. Using a newly-built cyclotron, Fermi was able to produce beams of charged pions with a maximum energy of a couple of hundred MeV and to scatter them from protons in the form of a liquid hydrogen target. The results of an equivalent modern experiment are shown

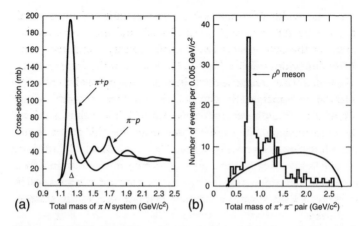

Figure 5.2 (a) Pion-nucleon cross-section as a function of the total effective mass of the πN system

(b) Plot of the number of events where the total effective mass of the two pions produced in the reaction $\pi^- + p \rightarrow p + \pi^+ + \pi^-$ has a specific value

in Figure 5.2a. This shows the 'total cross-section' as a function of the total effective mass of the pion–nucleon system, which includes the energy of their motion. (The total cross-section is the effective area of the target 'seen' by a particle in the beam and is proportional to the probability of any interaction. It is measured in millibarns (mb), where 1 barn is 10^{-28} m^2. There is a very pronounced peak in both the cross–sections where the interaction is particularly strong, indicative of the formation and subsequent decay of a resonance, which we now call the delta.

How do we know that this particle decays via the strong interaction when its expected lifetime would be around 10^{-23} seconds, far too short a time interval to be measured directly by even the best electronic methods? The plot is not unlike the spectral lines exhibited by excited atoms and nuclei. In those cases, the width of a line is measured in units of energy and denoted by ΔE, where Δ means 'small' and indicates that the width of the state is a relatively small quantity compared to its mass. The width is related to the lifetime, t, of the excited state, whose decay has given rise to the line, drawing upon Heisenberg's uncertainty principle. Specifically, the product of width of the resonance ΔE and t cannot be less than $\hbar/2$. The value of ΔE in the pion–nucleon case can be found from the plot and is about 100 MeV. So the lifetime is about 10^{-23} seconds, which indicates a particle decaying via the strong interaction. It is more usual to quote the width rather than the lifetime for a resonance if it decays by the strong interaction, because the width is easily measured.

The delta resonance, denoted by Δ (which has nothing to do with the use of this symbol in the uncertainty principle!), exists in four charged forms, represented as Δ^{++}, Δ^{+}, Δ^{0}, Δ^{-}, where for example the superscript '$++$' means a positive electric charge of two units. Subsequent analysis of its decay products found that the Δ has a spin of $\frac{3}{2}$, in units of $\hbar$, the first particle discovered to have a spin greater than one.

Searching for resonances by looking for enhancements in the final states produced in collisions of two particles has been a standard technique ever since Fermi's original experiment and has led to the identification of numerous hadron resonances. The 'golden age' of resonance discovery occurred during the mid- to late 1960s, as computers began to handle the large quantities of data being produced from a new generation of accelerators. As another example, this time for a meson resonance, Figure 5.2(b) shows the total effective mass of the two pions produced from a pion and a proton ($\pi^{-} + p \rightarrow n + \pi^{+} + \pi^{-}$). The curve shows the expected distribution of events if no resonance was produced. There is clear evidence for an enhancement at a mass of 760 MeV/c^2, which in this case is due to the production and decay to a resonance called the rho (ρ) meson, with a spin of 1.

Composite models and the Eightfold Way

With the rapid proliferation of new particles, including resonances, it was natural to seek a unifying model that would
explain these states and their quantum numbers. Several people
tried to interpret the known hadrons in terms of a smaller
number of elementary states. This was not new. Before the
discovery of strange particles, Fermi and the Chinese-American
theorist Chen-Ning Yang had proposed a model of the pion
as a composite bound state of a nucleon and an anti-nucleon.
Although they could explain some of the properties of the pion,
their model attracted very little interest until the discovery of
strange particles. At that point the Japanese theorist Shoichi Sakata
extended the idea to include these particles by taking as his
fundamental set of states the nucleons and the lambda, together
with their anti-particles. For example, in his model a kaon is a
composite of a lambda particle and an anti-nucleon, referred to as
$K = (\Lambda \bar{N})$, with the nucleon chosen to give the correct charge
on the kaon. One prediction of Sakata's model was that there
should be eight light mesons, although at that time only seven
were known (three charged pions, two kaons and the kaons' anti-
particles). When, in 1961, another neutral state, later called the
eta (η) meson, was discovered, it confirmed Sakata's prediction.

The model was less convincing when it came to explaining
baryons. When Sakata developed his approach, most physicists
believed that theories should put all hadrons on an equal footing –
'particle democracy' – a view that had arisen partly as a result
of the failure to produce a quantum field theory for the strong
interaction. Sakata's model was not democratic; some baryons
were singled out as being special. For example, in the model
a sigma hyperon is a composite of a lambda, a nucleon and
an anti-nucleon, $\Sigma = (\Lambda N \bar{N})$. But the sigma and lambda are
both hyperons with similar properties and there seemed to be no

reason why Sakata's choice should be preferred. (The objection that all particles were not being treated equally was rather ironic, because Sakata and his group wholeheartedly embraced Marxist philosophy.)

Around the same time, in 1960 and 1961, Gell-Mann and the Israeli theorist Yuval Ne'eman made a significant breakthrough in understanding the observed spectrum of hadrons. They considered the strangeness and charge quantum numbers of the spin-$\frac{1}{2}$ baryons. When plotted on the diagram shown in Figure 5.3a, they saw that the eight states of this *multiplet* form a hexagon. If the same is done for the spin-0 mesons, the pions and the kaons, as shown in Figure 5.3b, a hexagon also emerges, except at that time a particle was missing from its centre. For consistency across these two patterns, there would have to be an additional neutral meson. The η particle, which was found not long after Gell-Mann and Ne'eman constructed these diagrams, fits the bill. Throughout his career, Gell-Mann has had a knack for producing memorable names for theories and phenomena. In this case he named these patterns of eight the *Eightfold Way*, an allusion to the Eightfold Path of Buddhism.

Gell-Mann and Ne'eman's mathematical theory led them to expect that baryon multiplets could also be positioned naturally in a diagram, in this case in a regular decuplet pattern of ten constituents. At the time this plot was first made, there were nine baryon resonances, all with spin values of $\frac{3}{2}$, of which the delta was one. These are plotted in Figure 5.3c. Again they predicted the internal quantum numbers for the missing particle in the diagram. Gell-Mann also predicted its mass, since he had recognised that mass increased by about 150 MeV/c^2 every time the strangeness quantum number decreased by one. These properties indicated that the missing particle was a negatively charged baryon, which he named the omega-minus (Ω^-), with $S = -3$ and a mass of 1680 MeV/c^2. Gell-Mann's leap is often likened to the gaps that Mendeleev built into his periodic table of the chemical

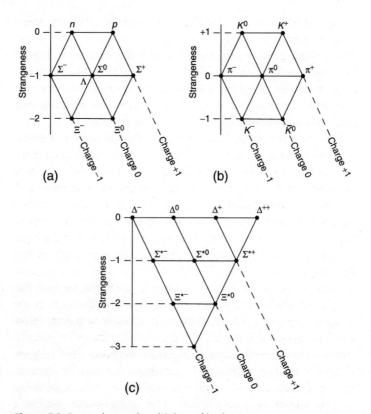

Figure 5.3 Some observed multiplets of hadrons

elements – gaps that were later filled when gallium, germanium and scandium were discovered.

In the Gell-Mann–Ne'eman decuplet diagram, all states except the omega-minus decay via the strong interaction or, in the case of the neutral sigma state, via the electromagnetic interaction, with characteristic decay times. In each case there is a ground-state baryon with the same value of S, so that strangeness can be conserved in the decay. For the omega-minus, however, there is no ground-state baryon with $S = -3$. Neither is there any lighter multi-particle state with $S = -3$ that would be allowed by conservation of energy, so the decay has to be via the weak interaction. Because of the selection rule that strangeness can only change by at most one unit at a time, the omega-minus was predicted to give rise to a multi-stage process of decays.

To test these ideas, an experimental team at Brookhaven National Laboratory in New York State used a hydrogen bubble chamber to look for a particle with the predicted quantum numbers and decay signature of the omega-minus. They found a match at almost exactly the predicted mass. The particle's production and decay is recorded in the remarkable picture in Figure 5.4, which shows an incoming negatively-charged kaon interacting with a proton and producing an omega-minus and a positively charged kaon. The Brookhaven team was confident the charges of the particles were correctly assigned, because the bubble chamber was in a uniform magnetic field that curved the tracks of charged particles. Because this is a strong interaction that conserves strangeness, they deduced that the other particle produced was a neutral kaon. The production process ($K^- + p \rightarrow \Omega^- + K^+ + K^0$) is shown in the right-hand diagram. The neutral kaon does not leave an ionisation trail as do charged particles, so it is shown as a dotted line. The omega-minus travels a short distance and then decays via the weak interaction (that is, $\Omega^- \rightarrow \Xi^0 + \pi^-$), where the strangeness quantum number changes by one. Similarly, the neutral xi is shown as a dotted line. Next, the neutral xi decays to two neutral particles, one a lambda and the other a neutral pion, neither of which leave a trail. They are detected by the weak decay of the lambda to a proton and a charged pion ($\Lambda \rightarrow p + \pi^-$) and the fact that the neutral pion decays to two photons, both of which produce electron-positron pairs. In these reactions and decays, electric charge and baryon number are also conserved. This figure gives a sense of the skill

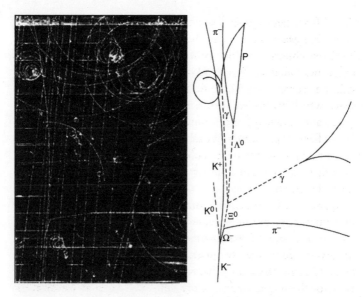

Figure 5.4 Characteristic pattern of tracks in a hydrogen bubble chamber from the production and decay of an Ω^- resonance

required to identify complex events when scanning thousands of bubble chamber photographs.

The quark model

Inspired by the Eightfold Way, in 1964 Gell-Mann and the young American theorist George Zweig simultaneously and independently realised that the observed pattern of particle multiplets would arise naturally if all the known hadrons were composites of three varieties of still smaller entities. Although they arrived at this conclusion by very different paths, their results were the same. Zweig called these constituents *aces*; Gell-Mann called them *quarks*, quoting the phrase 'Three Quarks for Muster

Mark!' from James Joyce's seldom read (and largely unreadable) book *Finnegan's Wake*. It has been said that the quarks refer to the three children of the main character in the book, who is sometimes called Mr Mark and is therefore presumably associated with the proton. Gell-Mann simply said he liked the sound of the word, which reminded him of a duck. The three quarks are the up (u) and down (d) and a new 'strange' quark, denoted by s. The collective properties of each quark type are said to define its 'flavour'. As suggested by its name, the strange quark was assigned a strangeness quantum number of $S = -1$, with the other two quarks having $S = 0$.

In the quark model, baryons are assumed to be clusters of three quarks qqq, bound together by the strong interaction, where initially q is any of the quarks u, d or s. Immediately, some unexpected properties of quarks emerge. First, they have non-integer electric charges and baryon numbers. The electric charges are $\frac{2}{3}$ for the u quark and $-\frac{1}{3}$ for the d and s quarks (in units of the magnitude of the charge on the electron) and the baryon numbers are $\frac{1}{3}$ for all three quarks. The finding for the electric charge was totally unexpected and seemed directly to contradict the results of classic experiments performed between 1910 and 1913 by the American physicist Arthur Millikan. He had shown that the smallest electric charge was that carried by a single electron and that other particle charges were integer multiples of this charge. But if we ignore this objection for the present, it is straightforward to work out the internal quantum numbers of all possible combinations of these three quarks. The result is shown in Table 2, which shows the possible baryon quark clusters in terms of their strangeness (S) and electric charge (Q). In the case of the decuplet of states with spin $\frac{3}{2}$, there is again an exact match between the ten predicted states and those observed experimentally. The same is true for the octet of states with spin $\frac{1}{2}$, provided we exclude clusters where all three quarks are the same.

Table 2 Predicted quark multiples for hadron clusters involving u, d, and s quarks

	Baryons					Mesons		
	S	Q				S	Q	
uuu	0	2	Δ^{++}		$u\bar{s}$	1	1	K^+
uud	0	1	Δ^+	p	$d\bar{s}$	1	0	K^0
udd	0	0	Δ^0	n	$u\bar{d}$	0	1	π^+
ddd	0	−1	Δ^-		$\left.\begin{matrix} u\bar{u} \\ d\bar{d} \\ s\bar{s} \end{matrix}\right\}$	0	0	π^0, η^0, η'^0
uus	−1	1	Σ^{*+}	Σ^+	$d\bar{u}$	0	−1	π^-
uds	−1	0	Σ^{*0}	Σ^0, Λ	$s\bar{u}$	−1	−1	K^-
dds	−1	−1	Σ^{*-}	Σ^-	$s\bar{d}$	−1	0	$\bar{K}^0$
uss	−2	0	Ξ^{*0}	Ξ^0				
dss	−2	−1	Ξ^{*-}	Ξ^-				
sss	−3	−1	Ω^-					

Mesons in the quark model are composite states of a quark and an anti-quark $q\bar{q}$, where q can be any of the quarks u, d or s. It is then simple to construct all possible meson states, as with the baryons. In this case, nine states (a nonet) are predicted, a contradiction of Gell-Mann's elegant Eightfold Way. However, the ninth state, called eta-primed (η'), was later discovered and the nonet was confirmed.

An equally impressive and significant fact about these predictions, both for mesons and baryons, is that there are no known states that do not follow from the simple hypotheses of the quark model. For example, there are no known mesons with strangeness quantum number $S = -2$ or baryons with $S = +1$.

What about spin? If baryons are made of three quarks, it follows that quarks must be fermions. Since spins can be added or subtracted provided the sum is not negative, three quarks, each with spin $\frac{1}{2}$, can have a total spin of $\frac{1}{2}$ (where two add and the third subtracts) or $\frac{3}{2}$ (where all the spins add). Baryons with spin-$\frac{1}{2}$ correspond to particles of the octet, which in the context of the quark model is the ground state multiplet, with the lowest possible energies for a given quark content. Baryons with spin $\frac{3}{2}$ correspond to the decuplet of resonances. Two observed particles with the structure *uds* appear in the spin-$\frac{1}{2}$ octet because, unlike the cases where at least two quarks are of the same type, the exclusion principle does not restrict the composite particle when each of the three quarks is a different flavour. For the baryon, there are two ways that the spins of the three quarks can give an overall value of $\frac{1}{2}$. In the first of these, the spin of a pair can be 0, which when combined with the spin of the third quark gives spin $\frac{1}{2}$ overall; in the second, the spin of the first pair can be 1, which when combined with the spin of the third quark can yield spin $\frac{1}{2}$ or $\frac{3}{2}$ overall.

For mesons, the total spin of a quark and an anti-quark (which also has spin $\frac{1}{2}$) can be 0 (where the two spins subtract) or 1 (where they add). The former corresponds to the ground state nonet and the latter to a nonet of resonances with spin 1, of which the rho meson shown in Figure 5.2b is an example. All the other members of the spin-1 resonance multiplet have been seen experimentally. Again, no states are predicted that do not occur in nature and no states are observed that are not predicted by the simple quark model.

By analogy with the electrons in atoms, quarks can have orbital angular momenta in addition to their spin. This would increase their energies (as it does with electrons in atoms) and lead to heavier hadron resonances with a range of higher spins, such as those seen in the pion-nucleon cross-sections of Figure 5.2a. Whenever spectroscopic predictions for resonances with higher spins have been tested, they have been confirmed by experiment.

Despite its early successes, the quark hypothesis was initially viewed with considerable scepticism because of the failure to observe free quarks. Most physicists looked upon quarks as a convenient mathematical description, rather than as actual physical particles. From his writings at the time, Gell-Mann himself was strongly inclined towards this view, although he downplayed this in later years. On the other hand, Zweig, who had taken a more intuitive approach in arriving at a theory of quarks, openly stated his belief that they were real particles. He paid a high price for his stance. When he submitted a paper to the leading American journal, it met with so much opposition from referees that he eventually withdrew it in exasperation. It was not fully published until sixteen years later. Gell-Mann, anticipating such problems, had prudently submitted his paper to a less prestigious journal.

The simple quark model seems to explain the world of particles extremely well. A baryon with the structure $qqqq\bar{q}$ would, for example, have the same internal quantum numbers as one with the structure qqq because the quantum numbers of the extra quark-anti-quark pair would cancel out. Similarly, mesons with structures of the form $q\bar{q}q\bar{q}$ would have the same internal quantum numbers as those with the structure $q\bar{q}$. Occasionally, experiments have claimed evidence for hadrons with these more complex structures but the observations have never been verified. Scientists conclude that only the particles of the simple quark model exist, although this does not answer the question of why the model works so well.

Colour

In the early years of the quark model, one of the simplest ways to rebut it was to note that quarks had not been seen. Doubters also had another line of attack: there was a conundrum embedded in the proposed structures of the model. For a wide range of states, the quark model explains the observed internal quantum numbers such as strangeness and the dynamic quantum number that is spin, but we have not considered the consequence of combining them. This is done by applying the exclusion principle, which states that in a collection of fermions, no two particles of the same type can have the same set of quantum numbers. Since quarks are fermions, the principle applies to them.

The problem this raises is most easily seen by considering the spin-$\frac{3}{2}$ baryons shown in Table 2. For example, the omega-minus baryon is a cluster of three s quarks, because this is the only combination that gives $S = -3$. Thus all three quarks have the same internal quantum numbers. But because the omega-minus has spin $\frac{3}{2}$, the quark spins must be oriented in the same direction, so that they add. So, all three quarks are identical and the exclusion principle is violated. This is not just a peculiarity of the Ω^- but is true for all baryons. For example, the doubly-charged delta resonance (Δ^{++}) is composed of three u quarks with aligned spins. Like the Ω^-, this state should not exist!

This is why when discussing spin-$\frac{1}{2}$ baryons, I excluded configurations where all three quarks were of the same type. To obtain the correct spin for the baryon, two of the quarks must have their spins aligned and the third quark spin must be in the opposite direction. Excluding these configurations is equivalent to allowing only those configurations where all three quarks have their spins aligned; that is, only allowing states that violate the exclusion principle. However, there is overwhelming evidence for the exclusion principle from atomic physics, so another solution had to be sought.

It did not take long before the American physicist Oscar 'Wally' Greenberg found the solution by resorting to the old device of particle physicists: invent a new quantum number. He called it, rather whimsically, *colour*, and hypothesised that quarks exist in three distinct colour states, which we now call red, blue and green, following the three primary colours of light.

Similarly, he assigned 'anti-colours' – anti-red, anti-blue and anti-green – to anti-quarks. Needless to say, colour in this context has absolutely nothing to do with the colours we experience visually. Greenberg further suggested that the quarks in hadrons are in different colour states: baryons would contain one red, one blue and one green quark. Then, because the three quarks would not be truly identical, the exclusion principle would not be violated! Pursuing the analogy with 'real' colour, combining the three primary colours of light yields white light, so the observed baryons are referred to as being 'colourless'; mesons consist of combinations of a coloured quark and an anti-quark carrying an anti-colour, so their observed states are also colourless. Greenberg's proposal that observable states have to be colourless is called *colour confinement*. It immediately explains why clusters of quarks such as $qqq\bar{q}$ and qq and other particles with fractional charges, are not observed experimentally. However, colour does not explain why states such $qqqq\bar{q}$ do not seem to exist.

What is colour, other than a convenient device to restore agreement with the exclusion principle? In 1972, the German theorist Harald Fritzsch, with his Swiss collaborator Heinrich Leutwyler and also Gell-Mann, hypothesised that colour plays the role in the strong interaction that electric charge plays in the electromagnetic interaction. As with the 'like charges repel, unlike charges attract' rule for electric charge, a colour 'charge' and its anti-colour would attract, yielding stable clusters of pairs of quarks, in which one has a definite colour and the other its anti-colour: these are the mesons. At the same time, states consisting of quarks with three distinct colours would also exist because of the mutual attraction of the three quarks of different colours: these are the baryons.

By the mid-1960s, particle physicists agreed that the simple quark model, incorporating the new quantum number of colour, gave a good account of the observed spectrum of hadrons, including both their ground states and their resonances.

When strangeness was introduced to explain the production and decay data for the kaons and hyperons, other predictions were made that were later verified by experiment, even though the reason for the existence of strangeness remained unclear. Colour, however, had been introduced purely to avoid violating the exclusion principle and so it was essential that it be tested in another context. In the next chapter we will see how the answers to these perplexing questions about quarks were resolved.

6
QCD, jets and gluons

After the introduction of colour, it became crucial to find direct evidence for the existence of quarks. This was obtained from experiments that scattered leptons from protons and others that studied the annihilation of electrons and positrons. Confirmation of the existence of coloured quarks enabled a practical quantum field theory of interacting quarks to be constructed.

Direct evidence for quarks

The success of the quark model in explaining the hadron spectrum was strong indirect evidence for the existence of quarks, but physicists still wanted to see some direct proof, ideally in the form of free quarks. Initially it was thought that if enough energy were brought to bear on hadrons, the strong force could be overcome and free quarks would emerge, something like ionising an atom or splitting a nucleus.

Experimenters tried colliding high-energy projectiles with protons. All that happened was that more particles, such as pions, were produced: no quarks were dislodged. Some physicists argued that perhaps the quarks were very heavy and that accelerators did not have sufficient energy to overcome the strong force. They suggested that in this case free quarks might be present in collisions induced by cosmic rays, some of which were known to have energies far higher than any man-made accelerator had achieved. Since quarks had to have non-integer electric charges, free quarks would necessarily be stable, because there are no lighter particles with non-integer charges to which they could decay. As a result, some bizarre searches were conducted of places where quarks might have come to rest, including deep-sea sludge and crushed oyster shells. Still no free quarks were found. Physicists were puzzled; conferences featured debates about the 'meaning' of quarks. According to his biographer, Gell-Mann tired of such discussions and got a doctor friend to write a fake medical note saying that he could no longer contribute to these talks because philosophy was bad for his health!

The first direct evidence for quarks finally came in 1968 in experiments at the SLAC laboratory in California, which essentially repeated the experiments conducted fifty years earlier that had discovered the existence of the nucleus but used different projectiles with higher energies. Rutherford, Geiger and Marsden had investigated the charge distribution within an atom by firing a beam of alpha particles at gold nuclei and observing the distribution of the scattered particles. Because of the relationship between distance and energy, probing the structure of hadrons requires projectiles with much higher energies – in the GeV region – than those used in that early experiment. SLAC had the ability to reach these higher energies. The team first used electrons scattered from protons, in a type of experiment referred to as *deep inelastic scattering*, because the projectiles, instead of scattering elastically, probe deep into the structure of the proton. The reactions studied are depicted in symbols as $e^- + p \rightarrow e^- + X$, where X stands for the many hadrons that were produced. Later experiments used muons as projectiles.

A study of the distribution of the scattered leptons, again comparable to what Rutherford and his colleagues had done, quickly showed that it was not consistent with the proton having a uniform charge distribution. Instead, the proton contained three much smaller entities. These were initially called *partons*, since the existence of quarks remained controversial. To prove that partons were actually quarks, experimentalists needed to find the values of their quantum numbers and show that they were the same as those predicted for quarks, specifically, spin $\frac{1}{2}$ and non–integer electric charges. To do this it was necessary to use neutrinos, which are uncharged and only interact by the weak interaction, as projectiles. Deep inelastic scattering experiments were carried out at a number of laboratories over the following years using ingeniously produced neutrino beams. Combining data from charged and neutral lepton scattering experiments unambiguously proved that partons did indeed have the same properties as quarks.

Feynman diagrams depicting the reactions studied are shown in Figure 6.1; the left panel depicts the use of electrons as the projectiles and the right panel the use of neutrinos. (The reason for the subscript μ on the neutrino will be explained in the next chapter.) In the former case, the probing particle is the exchanged photon and in the latter case it is a W boson, because neutrino scattering is a weak interaction. The target proton (entering from the left) is shown as consisting of three quarks; the exchanged particle interacts with a single quark, while the other two quarks take no part in the interaction – they are said to be *spectators*. This is similar in process to the beta decay of a nucleus, where one nucleon converts to another nucleon and all the other nucleons remain unchanged. In both cases, the scattered quarks and the

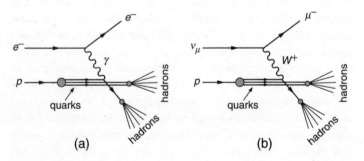

Figure 6.1 Deep inelastic scattering of electrons and neutrinos in the quark-parton model

spectator quarks contribute to the production of hadrons in the overall reaction, a process called *hadronisation*. This way of looking at the reaction, originally called the parton model but now more usually the quark-parton model, explains the data well and is an example of spectator models, where the interaction is with a single quark.

The evidence that there were quarks in nucleons was not totally unexpected. Nevertheless, the experiments provided a number of surprises. First, the quarks, far from being very heavy, were rather light, less than a third of the mass of the proton. At this point you may wonder what is meant by 'mass' when there are no free quarks to 'weigh'. It is worth taking a moment to consider this question. Take, for example, the electron: in QED, an electron constantly emits and reabsorbs virtual photons and these photons constantly create and reabsorb pairs of virtual electrons and positrons, and so on. What is actually measured when scientists 'weigh' an electron is an effective mass that includes the contributions to the total energy from this cloud of particles. These are examples of virtual particles – particles that do not appear in either the initial or final states as seen in the laboratory – that were introduced when I discussed Feynman diagrams previously. The exchanged photon and W boson in Figure 6.1 are also virtual particles. In the case of quarks, the effective mass includes the effects of the cloud of the equivalent force carriers of the strong interaction and the sea of virtual quark-anti-quark pairs they in turn produce. For the quarks in nucleons, the u and d quarks, these contributions are not small. Thus, in quoting masses for quarks we can either give the 'effective' mass as deduced from studies of hadrons, usually called the 'constituent' mass, which is what I have done above, or the theoretical mass without these contributions. The latter is much smaller, just a few MeV.

The deep inelastic scattering experiments also showed that as the energy of the projectiles increased, so that they probed deeper into the structure of the proton, the quarks unexpectedly behaved as if they were free particles trapped within the proton. Presumably the same is true for all hadrons. Finally, if the proton consisted of just three quarks, it would follow that the sum of the momentum of these quarks would have to be equal to that of the proton, and for a fast-moving proton, each quark would have approximately one-third of the proton's momentum. Also, against expectations, this proved not to be the case and an individual quark momentum was found to vary from event to event. Moreover, the sum of the constituent quarks' momenta did not add up to the proton's momentum. Quarks were found to have only about half of the proton's momentum, so a substantial fraction of the momentum was being carried by other particles: these turned out to be gluons.

Direct evidence for colour

Even though direct evidence had been obtained for the existence of quarks in hadrons, there remained the question of colour. The key piece of evidence for colour came from experiments at e^+e^- colliders. When electrons and positrons collide, they annihilate to produce a single virtual photon, which can then produce pairs of real matter and anti-matter particles, including muons and anti-muons and quarks and anti-quarks of any flavour, provided energy is conserved. The quarks and anti-quarks do not appear as free particles but cluster to produce various hadrons depending on the energy of the initial particles. (In symbols, these two reactions are $e^+ + e^- \rightarrow \gamma \rightarrow \mu^+ + \mu^-$ and $e^+ + e^- \rightarrow \gamma \rightarrow q + \bar{q}$.) Both are electromagnetic processes and at high energies the main difference in them is the charge of the particles in the final states. The rates for these reactions depend on the squares of the electric charges involved, so the ratio R of the production rate of $q\bar{q}$ pairs (which, because of hadronisation, is a measure of the production rate of hadrons) to the rate for production of $\mu^+\mu^-$

pairs is equal to the sum of the squares of the electric charges on the quarks. Assuming the initial leptons have only sufficient energy to produce the u, d and s quarks and their anti-particles and using the hypothesised values of their electric charges, this ratio was predicted to be $\frac{2}{3}$. However this prediction completely ignores colour. If colour were a real property of quarks, each quark and anti-quark could be produced in each of the three different colour states with equal probability. This follows because although the photon is exchanged between charged particles (it is said to 'couple' to charged particles) it is 'blind' to colour. So the prediction would need to be multiplied by 3, the number of colours, which would give a ratio of 2.

The first experiment to test the existence of colour was conducted in 1970 and data from later experiments considerably reduced remaining uncertainties. Across the board, the findings ruled out a ratio of $\frac{2}{3}$; they were consistent with a ratio of 2. This unambiguous existence for the colour quantum number convinced the few remaining doubters about the reality of quarks and set the stage for the construction of a quantum field theory of the strong interaction.

Quantum chromodynamics

The quantum field theory of interacting quarks was constructed by analogy with the theory of electromagnetic interactions, QED. It was assumed that quarks interact via the exchange of massless spin-1 particles, which came to be known as gluons. The source of the interaction is the colour charge, just as the source of the electromagnetic interaction is the electric charge. For this reason, the theory is called quantum chromodynamics (QCD), yet another name coined by Gell-Mann.

There are many similarities between QED and QCD. Both are theories of forces being transmitted by spin-1 bosons and both have the property of gauge invariance. Because of the work of t Hooft, QCD is known to be renormalisable, so not plagued by the infinities that defeated earlier attempts to construct quantum field theories. There is also a parallel with the electromagnetic forces seen in certain energy levels in atoms and in the strong forces seen between quark clusters. For example, in atoms, in addition to the interaction between the electric charges of the electrons and the protons, there is also an interaction between the intrinsic magnetic moments of the electrons and those of the protons and neutrons that produces small shifts in the energy levels of atoms.

Why does the electrically neutral neutron even have a magnetic moment? The answer lies in the fact that the sign of a particle's magnetic moment depends on the orientation of its spin. The neutron is a composite object made of three quarks and although the charges on the quarks must cancel exactly to give zero electric charge for the neutron, the magnetic moments of the three quarks do not, because their three spins must combine to give the correct spin of the neutron. The three spins cannot all be in the same direction.

In hydrogen, the simplest atom, the magnetic interaction changes the energy level by a very small amount, which can be positive or negative depending on the orientations of the spins of the electron and the proton: positive for total spin 1, negative for total spin 0. A similar effect is seen in hadrons, due to the magnetic interactions of their constituent quarks. For example, the octet of spin-1 resonances, including the rho meson, is heavier than the octet of spin-0 ground states (shown in Figure 5.3). The effect is much larger in mesons because they are vastly smaller than atoms and this is further enhanced by the fact that the strong interaction between quarks is far stronger than the electromagnetic interaction. Whereas the numerical value of the electromagnetic coupling constant α (also known as the fine structure constant), is approximately $\frac{1}{137}$, the value of the strong interaction equivalent, denoted α_s, is approximately forty times larger.

Although there are many similarities between QCD and QED, there are also profound differences. First, although a photon couples to charged particles, it is not itself electrically charged, so photons do not couple directly to other photons. In QCD, the equivalent particles are gluons and they couple to quarks that carry the colour quantum number. Just as electric charge has to balance in interactions, so does colour charge. If a quark of one colour (for example, red) changes to a quark of a different colour (for example, blue), the exchanged gluon must carry a combination of colours (in this case, red-blue) to balance the colour charges. So gluons carry a colour quantum number and do couple directly to other gluons. By considering all possible transformations of quarks of a definite colour, it is possible to deduce that there are eight possible combinations of colour that can be carried by gluons, that is, there are eight distinct varieties of gluons. QCD is a more complicated theory than QED, which has just two electric charges, positive and negative and only one force carrier, the photon. It is worth emphasising that in the strong interaction, colour determines the strength of the force, not the flavour of the quark that is carrying colour. The strong interaction is flavour-independent, just as the nuclear strong force is independent of electric charge.

The fact that gluons couple directly to other gluons raises the interesting possibility that, provided the forces were strong enough, particles consisting of colourless combinations of gluons that would not violate colour confinement could exist. These have been given the name *glueballs*; several experiments have been conducted to search for them. The obstacle to finding glueballs comes from the difficulty in predicting the masses of particles using QCD. What predictions there are suggest that the forces between gluons could be strong enough to support stable clusters. Unfortunately, the calculations also suggest that if glueballs exist, their masses fall in a region where several standard, electrically-neutral meson resonances – clusters of a quark and an anti-quark – already exist. It has proven problematic to distinguish possible glueballs from these other states experimentally. To date there is no definite evidence for the existence of glueballs but some slight evidence for states that could be clusters of quarks and glueballs – the so-called *hybrid mesons*. If glueballs exist, they would help to expand our understanding of QCD.

The simple fact that gluons couple directly to other gluons creates a crucial difference in the behaviour of QCD compared to QED. In QED, although the electromagnetic coupling α is referred to as a coupling constant, this is not the whole story. Instead, as a consequence of the cloud of virtual particles that always surrounds a charged particle, the coupling constant increases slightly as energies increase or, equivalently, distances decrease. The cloud essentially screens the charge of the particle and so changes its effective charge as seen by a second charged particle. This behaviour is an intrinsic property of QED. In QCD, a similar cloud of virtual particles exists, which consists of gluons and quark–anti-quark pairs. However, because of the gluon–gluon interactions in QCD, it turns out that this produces an 'anti-screening' effect and the coupling for the strong interaction α_s decreases with energy. Because of these energy dependencies, the couplings are often referred to as 'running' coupling constants.

The behaviour of the strong interaction coupling gives a qualitative explanation for why quarks in hadrons appear to behave as free particles when they are probed with high-energy projectiles: the coupling between the quarks decreases as energies increase, that is, as shorter distances are probed. This phenomenon is referred to as *asymptotic freedom*, since quarks would become free particles at infinitely large energies. Conversely, the extra interactions also explain why quarks are permanently confined in hadrons, because as the quarks are separated, the force between them increases. This is something like stretching an elastic string: as the length of the string is increased, a restoring force proportional to the extension comes into play and this force continues to resist the stretching until at some point the string breaks and the two parts of the string are freed (Hooke's Law of classical physics). In the case of QCD, the two quarks are not freed. Instead, as the force between the quarks increases, it resists quark separation until the energy generated by the gluon fields between the quarks becomes sufficiently strong to create new hadrons, yet the quarks remain confined. This property of QCD is referred to, rather dramatically, as *infra-red slavery*.

Experiments support this picture of coloured quarks interacting via the exchange of coloured gluons. One piece of evidence concerns the deep inelastic scattering experiments that gave the first direct evidence for the existence of quarks in hadrons. In the parton model used to interpret these experiments, it was assumed that the confined quarks had no mutual interactions. But in QCD it is necessary to take into account these mutual interactions via the exchange of gluons. There are two effects: first, the leptons not only interact with one of the three quarks comprising the proton but also with the sea of virtual quarks and anti-quarks surrounding it. Second, the energy dependence of the strong interaction coupling has to be taken into account. These QCD modifications produce small changes in the predictions for deep inelastic scattering data, which have been verified by more recent and precise experiments.

Another piece of evidence came from experiments conducted to test whether quarks possessed colour. The early experiments differentiated between two fixed ratios, depending on whether colour existed or not. Later experiments at higher energies revealed a small energy dependence in this ratio. This is illustrated in Figure 6.2, which shows the measured ratio R for experiments in the energy range 10 GeV to 40 GeV. At these energies, quarks that are heavier than the u, d and s quarks can be produced and the value of R is larger. As the quarks and anti-quarks accelerate away from the point where they are produced, they radiate photons (because the quarks are charged) as well as gluons, the

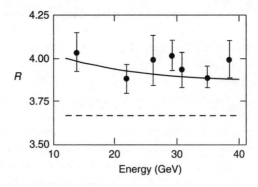

Figure 6.2 The ratio R of the cross-sections for hadron to muon production in e^+e^- annihilations

force carriers of the strong interaction. Radiation of photons can be calculated using the electromagnetic coupling α and radiation of gluons calculated using the strong coupling, α_s. These effects result in the energy dependence of R. In particular, the strong coupling varies significantly in the energy region under investigation. The solid curve in Figure 6.2 depicts the prediction including these corrections, while the dotted line is the prediction of the simple parton model. Both curves assume the existence of colour. Without colour, both predictions would have to be divided by a factor of three and would be in total disagreement with the data.

There are many more examples of coloured quarks interacting via the exchange of coloured gluons. Even in processes that cannot be calculated from first principles using the method of perturbation theory, QCD can be used to provide a detailed framework for experimental analyses. Whenever QCD has been tested, it has passed the test, with predictions that agree with experiment to accuracies better than one part in a thousand. Although this does not match the phenomenal precision of QED, it is still a very impressive vindication. Ideally, physicists would like a mathematical proof of confinement in QCD but just as for the calculations on glueballs, this has not yet been possible.

Jets and the direct evidence for gluons

Although QCD quickly proved to be a successful explanation of a wide range of experimental results, direct evidence for the existence of gluons still needed to be found. Unfortunately, just as for free quarks, it is not possible to detect free gluons; gluons are coloured and free gluons are forbidden to exist by the requirement of colour confinement. However, direct evidence for gluons emerged in 1979 from an e^+e^- annihilation experiment using an electron-positron collider at the Deutsches Elektronen-Synchrotron (DESY) laboratory in Hamburg, Germany, and was confirmed in later experiments at other colliders. Using electrons and positrons is more efficient than using interactions involving protons, since conservation of baryon number means that energy is always 'wasted' in producing baryons from proton targets.

When electrons and positrons annihilate, pairs of quarks and anti-quarks can be produced. As these move apart, the force between them increases in accordance with the predictions of QCD. This process continues until the strong gluon field has enough energy to create further pairs of quarks and their anti-particles, which quickly undergo hadronisation and form clusters that can be observed in the laboratory as hadrons. These annihilations have been studied for a range of energies up to more than 100 GeV. As the energy increases, QCD predicts that there should be an increasing tendency for the hadrons to be contained within a small angular region tied to the direction in which the original quarks were produced. In other words, the hadrons 'remember' the mechanism by which they were produced, as illustrated in Figure 6.3a. At high energies two prominent *jets* of hadrons should appear back-to-back, ensuring momentum conservation. Figure 6.3b shows an electronic reconstruction of the paths of particles in a track chamber from an e^+e^- collider at DESY. The view of the interaction region is taken along the beam pipe. As usual, the curvature of the tracks is due to an applied magnetic field. The two-jet events beautifully confirm the predictions of QCD.

What role do gluons play in this? As the quarks and anti-quarks accelerate away from the point of interaction they radiate gluons

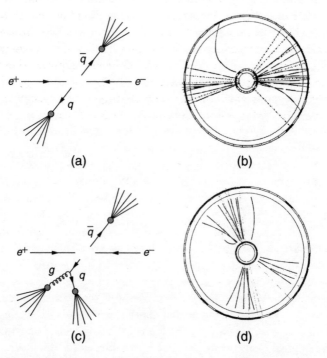

(a)   (b)

(c)   (d)

Figure 6.3 Two-jet and three-jet events in an e^+e^- collider experiment

but usually these carry little momentum, so any jets they generate will be subsumed within the quark jets. However, if a gluon is produced with a momentum comparable to that carried by a quark, to conserve momentum the direction of the quark will change and the jets produced by the gluon and quark will have distinct directions. Figure 6.3c illustrates this mechanism, where the gluon, g, is drawn as a coiled line. This effect occasionally happens. A typical 'three-jet' event, recorded at DESY in 1980, is shown in Figure 6.3d. This is probably the closest we will ever get to seeing a gluon.

By analysing their energies and angular distributions, we can show these three-jet events really are evidence for gluons. It is possible to deduce which jet is due to a gluon and moreover, the spin of the particle that produced it. The data are only compatible with spin 1, completing the direct confirmation of the existence of gluons. Further, because the ratio of the rate of production of three-jet events, those involving gluons, to the rate for two-jet events depends on the size of the strong coupling α_s, quantitative tests of QCD are possible. Again, QCD passes the test.

7
Weak interactions

The weak interaction often does not conserve quantum numbers that are conserved by the strong and electromagnetic interactions. This is true for strangeness and for other quantum numbers. The challenge is to understand how these differences arise and to give a systematic description of observed weak interaction scattering and decay processes. If we ignore gravity, the weak interaction is the feeblest of the fundamental forces at everyday energies; this makes it difficult experimentally to study scattering reactions, unlike the strong or electromagnetic interactions. Nevertheless, the long-awaited discovery of the neutrino in the mid-1950s launched the field of neutrino physics and led to a means of measuring neutrino masses. In this chapter, I will explain the basic ideas of the weak interactions of leptons and hadrons as they interact via the exchange of one of the force carriers of the weak interaction, the W boson.

Lepton multiplets and lepton numbers

There are three charged leptons: the electron and its two heavier siblings, the muon and the tauon. The muon was discovered in cosmic ray experiments and was initially mistaken for the pion, the particle predicted by Yukawa. Its mass is 106 MeV/c^2, about two hundred times the mass of the electron. The tauon, which was discovered in 1975 in electron–positron annihilation experiments, is much heavier, about 1777 MeV/c^2. The taon's properties have been much less studied than those of the two lighter charged leptons but all appear to be structureless elementary spin-$\frac{1}{2}$ particles.

Until now, I have used 'neutrino' when referring to an uncharged lepton but just as there are three charged leptons, there are three neutral ones: the electron neutrino, the mu neutrino and the tau neutrino, denoted ν_e, ν_μ, and ν_τ. For example, the electron neutrino (and its anti-particle) is involved in nuclear beta decays. How do we know the three neutrinos are distinct? One piece of evidence is that electron neutrinos produce electrons when they interact with neutrons ($\nu_e + n \rightarrow e^- + p$) but do not produce muons ($\nu_e + n \nrightarrow \mu^- + p$). Similarly, physicists observe that mu neutrinos produce muons $\nu_\mu + n \rightarrow \mu^- + p$ but not electrons, $\nu_\mu + n \nrightarrow e^- + p$. Clearly the two neutrinos are associated with their own charged lepton and not with others. The tau neutrino was only detected in 2000; no beams of these particles exist to perform similar tests, but there is no reason to suppose that its behaviour is any different from its lighter siblings. The six leptons are classified into three generations, each consisting of one charged lepton and its associated neutrino. Just as for quarks, particle physicists refer to the leptons in each of these three generations as having different flavours.

As we have seen before, if some reaction is not observed when there is apparently nothing to prevent it occurring, it is an indication that a conservation law is in operation. In the case of interactions involving leptons, the experimental data are explained by defining new *lepton numbers*. The electron and its neutrino are assigned an electron lepton number equal to 1, the muon and its neutrino are assigned a muon lepton number, also equal to 1, and for the tauon and its neutrino similarly. All their anti-particles have appropriate lepton numbers -1 and all other particles are assigned lepton numbers of 0. The three lepton numbers are distinct and separately conserved in all interactions. By adding lepton numbers on both sides of the reactions noted above, we can see that this is consistent with the observed and forbidden reactions. Many other reactions support the idea of lepton numbers being individually conserved. The decay of the negative muon can be verified by examining the neutrinos produced: electron neutrino anti-particles and mu neutrinos ($\mu^- \rightarrow e^- + \bar{\nu}_e + \nu_\mu$). Also, the muon does not decay via an electromagnetic process that yields electrons and photons ($\mu^- \nrightarrow e^- + \gamma$),which would lead to a lifetime much shorter than that observed. Despite extensive searches by experimentalists, reactions that do not conserve lepton numbers have never been observed.

Because neutrinos only interact via the weak interaction, very large masses of detector material are required to ensure that sufficient interactions occur. The electron neutrino (strictly, the anti-neutrino) was first detected in 1959 in an ingenious experiment carried out by two American physicists, Fred Reines and Clyde Cowan. At that time no neutrino beams existed, so they turned to a nuclear fission reactor as the source of neutrinos. Fission reactors produce a huge flux of electron anti-neutrinos from the beta decay of nuclei – as many as 10^{17} per square metre per second. Even so, Reines and Cowen only found about two anti-neutrinos an hour interacted with the protons in a very large detector to produce neutrons and positrons (via the reaction $\bar{\nu}_e + p \rightarrow n + e^+$). Their problem was to detect the two final particles, thus verifying the existence of the anti-neutrino, in a way that distinguished them from particles produced by other interactions that might produce 'spurious' signals.

A diagram of the apparatus used by Reines and Cowan is shown in Figure 7.1a. It consisted of two large tanks containing an aqueous solution of cadmium chloride as the target, sandwiched between three tanks of liquid scintillator as the detector, the whole surrounded by heavy shielding to prevent all particles, except for neutrinos and anti-neutrinos, from entering the device. The overall weight of the apparatus was about ten tonnes. When a reaction occurred, the positron rapidly annihilated with an atomic

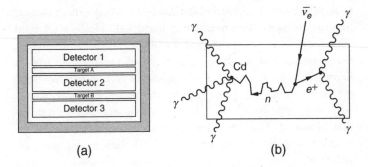

Figure 7.1 First neutrino detection experiment

electron and two photons were produced, which gave coincident electronic pulses in the two surrounding tanks of scintillator, that is, pulses occurring practically simultaneously. This is shown schematically in Figure 7.1b, which depicts an interaction in one of the target tanks.

In contrast, the neutron suffered a number of scatters from the protons in the water until it was slowed down sufficiently to be captured by the cadmium nucleus, a process that released more photons. These photons also produced coincident pulses in the same pair of scintillator tanks, both above and below the target. The crucial point is that the second pulses were delayed by the time it took for the neutron to be slowed, which Reines and Cowan calculated to be just a few microseconds. Overall, the reaction had a very distinctive signature; the chance that this signal would be reproduced by neutrons and positrons produced accidentally by some other processes was calculated to be quite low. The first run of the experiment, conducted over about three months, produced 567 events with the expected signals. Only 209 would have been detected if they had been due to 'accidents'. For the first time, anti-neutrinos had been detected.

Just as the detection of neutrinos is difficult, so is making measurements of their masses. Again, I will first consider the electron neutrino, since this particle has been the most studied. One way to measure the mass of the electron neutrino is from the beta decay of nuclei. The presence of a neutrino in the products of the decay implies that the electron has a continuous energy distribution. (This was one reason why Pauli originally postulated the existence of the neutrino.) The behaviour of this distribution near the point where the electron has its maximum energy depends on the mass of the neutrino. If the distribution can be measured very near this maximum point, in principle the mass of the neutrino can be found, but the closer one gets to the end point, the fewer decay events there are. In practice, therefore, experimentalists measure close to the end of the distribution and then extrapolate to the actual end point. The best result comes from the beta decay of the element tritium and suggests that the value of the electron neutrino's mass ($m_{\bar{\nu}}$) is less than 2 eV/c^2 or about four-millionths of the mass of the electron. The other two neutrinos do not take part in the beta decays of nuclei and their masses are much less well known. They are obtained by finding what values of the neutrino masses are consistent with energy conservation in the decays of the muon ($\mu^- \rightarrow e^- + \bar{\nu}_e + \nu_\mu$) and the tauon ($\tau^- \rightarrow e^- + \bar{\nu}_e + \nu_\tau$). This yields bounds at least 10^5 times larger than that for the mass of the electron neutrino.

In the standard model, the decays of the muon and tauon are mediated by the exchange of a W boson, as shown in the Feynman diagrams in Figure 7.2. Such reactions are called *weak charged interactions*, because the exchanged boson is electrically charged. From these diagrams it is possible to calculate the lifetimes of both particles. By comparing them with the experimentally measured values, physicists have deduced the intrinsic strength of the weak charge interaction, α_w. This coupling constant is a measure of the strength of each of the vertices in Figure 7.2, just as the electromagnetic coupling constant α is a

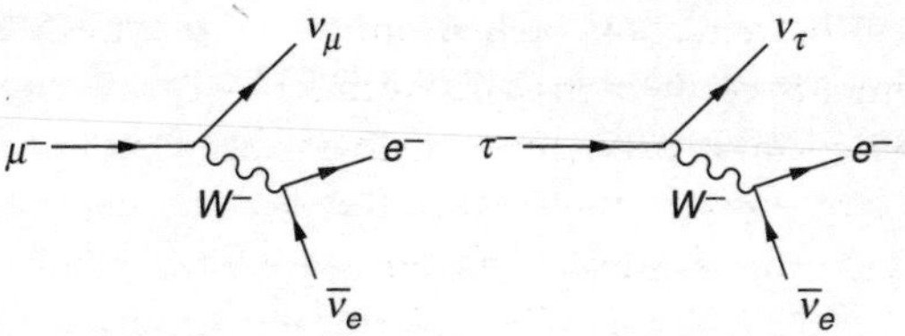

Figure 7.2 Feynman diagrams for muon and tauon decays

measure of the strength of the electromagnetic interaction. The value of α_w turns out to be the same for muon and tauon decays and is in fact not much different in size from the value of α. At first sight this is very surprising, since the muon lifetime is approximately 2×10^{-6} seconds, whereas that of the tauon is approximately 3×10^{-13} s, a factor of 10^7 smaller.

To make sense of this result, recall that particle lifetimes depend not only on the weak coupling constant but also on how much energy is released in the decay – the more energy released, the shorter the lifetime. Because the tauon is much heavier than the muon, its lifetime is much shorter. We have already seen an example of this in the decay of the free neutron, where, because the neutron-proton mass difference is so small, the lifetime of the neutron is measured in minutes rather than fractions of a second, as would be expected for a weak decay. The equality of the weak interaction couplings of muons and tauons to W bosons holds for all interactions involving leptons; it is called *universality*. It is odd that two generations of heavier leptons exist, since they do not seem to lead to any new information about the weak interaction that could not have been deduced from the first generation of particles.

Why does the weak interaction appear to be much feebler than the electromagnetic interaction despite the fact that their couplings are of a similar size? The reason can be traced to the different masses of the exchanged bosons. In the weak interaction, the mass of the W boson suppresses the intrinsic strength of the interaction. Because it is heavy, what is observed at everyday energies is much feebler than the electromagnetic interaction, even though their intrinsic strengths are similar.

Neutrino mixing and neutrino masses

The measurement of the electron neutrino mass from the beta decay of tritium does not rule out the possibility that neutrinos are massless, as the standard model initially assumed. Neither do the measurements of the masses of the other two neutrinos. This is an important question, because if neutrinos do have non-zero masses, some interesting effects can occur, with consequences elsewhere.

These phenomena stem from neutrino *mixing*. Mixing is a general feature of quantum theory and refers to the fact that a set of quantum states (in this case particles) can be expressed as linear combinations of an equivalent set of states. In the present context, mixing is the assumption that the observed neutrinos ν_e, ν_μ, and ν_τ do not have definite masses. Instead, they are linear combinations of three other neutrino states, ν_1, ν_2, and ν_3, that do.

Suppose a beam of neutrinos, ν_e, with a definite flavour, is produced in a reaction. At the point of production, each electron neutrino is a combination of the states ν_1, ν_2, and ν_3, each of which, because of their slightly different masses, has a slightly different energy. Since particles also behave as waves in quantum theory, each of these particles will have an associated wave with a slightly different wavelength and hence frequency. As the system evolves, the three waves of the initial beam of electron neutrinos oscillate as they travel through space. Because each component is itself a combination of the states ν_e, ν_μ, and ν_τ, at any instant the beam will contain varying amounts of all three flavours of neutrino. So, if the number of electron neutrinos were measured, we would find that some had 'disappeared' to be replaced by neutrinos of other flavours, a situation called *neutrino oscillation*. It is similar to the beats heard when sound waves of slightly different frequencies are superimposed. In quantum theory, it can only occur if neutrinos have non-zero masses. So neutrino oscillations allowed physicists in principle to measure neutrino masses.

The suggestion that neutrino oscillations occur arose from a long-standing problem in astrophysics, associated with the energy output of the Sun. Solar energy is derived from a sequence of nuclear fusion reactions and results in an enormous flux of low-energy electron neutrinos. These neutrinos were first detected at the surface of the Earth in an experiment begun in 1968, headed by the American astrophysicist Ray Davis. A huge tank of perchloroethylene (dry-cleaning fluid) was placed deep underground, to shield it from cosmic rays, in the Homestake gold mine near the town of Lead, South Dakota, USA. The neutrinos interacted with the chlorine in the fluid and converted it to an unstable isotope of argon with a lifetime of thirty-five days. This is a weak interaction; only a single argon atom was produced every few days. These atoms were flushed out of the tank every few weeks and counted.

The expected counting rate for electron neutrinos came from a detailed model – the standard solar model – that had been built up over many years by astrophysicists to explain the nuclear reactions powering the Sun. However, using the data accumulated over twenty years (yes, twenty years!) of observations at the Homestake mine, Davis showed that the standard solar model prediction was totally incompatible with the results of his experiment – the observed number of electron neutrinos was only about one-third of the expected number. This discrepancy was dubbed the *solar neutrino problem*. Davis's consistent results, over such a long time, forced others to take seriously the suggestion that electron neutrinos from the Sun might have oscillated to neutrinos of other flavours on their way to Earth, and spurred experimentalists to design experiments to look for neutrino oscillations.

There are several types of oscillation experiment. Some start with muon neutrinos, others with electron neutrinos or their anti-particles; some study solar neutrinos, others use neutrinos produced in the atmosphere indirectly by cosmic rays and yet others use the neutrinos produced in nuclear reactors. The first experiment that proved the existence of neutrino oscillations was conducted by the Japanese SuperKamiokande group, which in 1998 studied atmospheric neutrinos. These neutrinos are produced from the decays of pions, which themselves are products of cosmic rays interacting with atoms in the upper atmosphere. The detector was a huge stainless steel cylindrical tank, about forty metres in diameter and forty metres high, filled with very pure water and situated 1000 m deep underground in the Japanese Alps. It was monitored by more than twelve thousand photomultipliers that collected the light emitted as Čerenkov radiation by fast-moving charged leptons produced in the tank when a neutrino interacted with a nucleon in the water. (The installation suffered a devastating blow in November 2001 when about 6600 of the photomultiplier tubes, costing about US$3000 each, imploded, apparently in a chain reaction as the shock wave from the concussion of each imploding tube cracked its neighbours.)

The decay sequence of a pion to a muon and its neutrino, followed by the decay of the muon to an electron and its neutrino, means that an initial pion should give rise to two mu neutrinos for every one electron neutrino. Instead of this expected ratio, the value measured by the SuperKamiokande group was about 1.3, suggesting that oscillations had removed some of the mu neutrinos. The experiment was also able to compare the flux from neutrinos coming into the detector from directly above to the flux from neutrinos coming from below. Since the flux of cosmic rays hitting Earth is the same in all directions, in the absence of oscillations these two rates should be equal. Time, however, changes everything. The downward-moving neutrinos have travelled a relatively short distance from their point of production in the atmosphere, whereas the upward-moving ones travelled through the whole diameter of Earth, providing plenty of time to oscillate, perhaps by several cycles. The group's measurements showed no difference in the fluxes for electron neutrinos, consistent with no oscillations, but for mu neutrinos the upwards flux was about a factor of two lower. This indicated that the mu neutrinos had oscillated, presumably to tau neutrinos. In 2004, the experiment was refined to measure the flux of neutrinos as functions of the distance they had travelled before reaching the detector. This explicitly demonstrated neutrino oscillations.

I have described the SuperKamiokande group's research in some detail because it was the prototype for a number of later experiments. Some have used neutrino beams produced at accelerators, while others have used solar neutrinos but with detectors that were capable of detecting lower-energy neutrinos that come from the main fusion reaction in the Sun. One of the latter experiments, carried out at the Sudbury Neutrino Observatory (SNO) at Sudbury, Ontario, Canada, in 2002, produced conclusive proof that oscillations are the solution to the solar neutrino problem. The experiment, sited in a mine about two kilometres underground, consists of an acrylic sphere, twelve metres in diameter, filled with heavy water (a form of water where the nucleus of hydrogen has an additional neutron) as the Čerenkov medium, surrounded by ten thousand photomultiplier tubes. Figure 7.3 shows the sphere, covered in photomultipliers, during construction. The very large size, needed because neutrinos have such a low probability of interacting, can be judged by the human figures on the gantry. The cavern, which measures thirty-four metres by twenty-two metres, was itself filled with very pure water to shield the sphere from naturally-occurring radiation from the rock walls and to

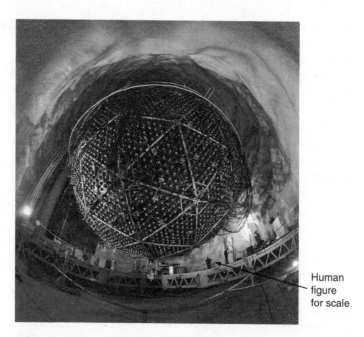

Human figure for scale

Figure 7.3 SNO neutrino detector during construction, showing the photomultipliers surrounding the acrylic sphere

reveal charged particles entering the heavy water from outside the sphere. The use of heavy water enabled the researchers to look at several reactions simultaneously. One of these established the correctness of the standard solar model and the others showed that a substantial fraction of the original electron anti-neutrinos emitted by the Sun had oscillated into other flavours. The research team predicted that such oscillations should also be observable in neutrinos obtained from nuclear reactors, as used in the original Reines and Cowan experiment, provided the distance from the reactor to the detector was at least one hundred kilometres. This was later confirmed by a team of Japanese physicists who detected neutrinos emitted from more than sixty fission reactors throughout Japan and South Korea.

Neutrino oscillations tell us a great deal about neutrino masses. To start with, oscillations only occur because the neutrinos have non-zero masses. However, finding values for neutrino masses is difficult, first because mixing involves differences in the squares of the masses of any pair of neutrino states ν_1, ν_2, and ν_3 and not the individual masses themselves and also because it contains some parameters that have to be determined from experiments. Finding neutrino masses requires a full set of all possible experiments that can be conducted; to date not all of these have been done. What has emerged so far are limits that define a range of values for the squared mass differences. By making some reasonable assumptions, particle physicists can use these to deduce that all three states, ν_1, ν_2, and ν_3, have masses that are almost certainly less than $2\ eV/c^2$.

Neutrino oscillations might seem to have some implications for lepton number conservation in weak interactions but in practice this is not the case. Consider the decay of a negatively charged tauon to leptons, as shown in Figure 7.2: in principle, the two neutrinos in this decay could oscillate to neutrinos of other flavours and so the observed decay would violate lepton numbers. However, because the W boson is very heavy, the range of the weak interaction is miniscule compared to a typical oscillation length, so the probability of an oscillation within that range is negligible. For that reason, lepton number conservation can still be used with confidence to identify neutrinos from reactions and decays.

Hadron decays

Most of the ground states of quark clusters decay by the weak interaction, which often produces leptons in the final state. In the

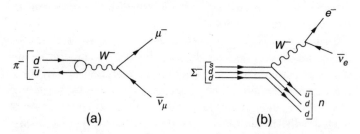

Figure 7.4 Feynman diagrams for decays of charged pions and sigma baryons

quark model, this means that quarks change from one flavour to another by interacting with an exchanged boson but are still confined within a hadron. Figure 7.4 shows examples of how this works for the decays of a charged pion and a charged sigma baryon. In the first case, the d and $\bar{u}$ quarks that constitute the negative pion π^- annihilate and form a virtual W^- boson, which then decays to a pair of leptons that is observed in the laboratory, giving $\pi^- \rightarrow \mu^- + \bar{\nu}_\mu$ overall. For the decay of the negative sigma hyperon Σ^-, the s quark transforms to a u quark by emitting a virtual W^- boson, which also decays to a lepton pair. The strange quark can change to a non-strange quark because strangeness is not conserved in these weak decays. Electric charge is conserved at both vertices in each diagram. In the spectator model, only one quark takes part in the interaction. The other two spectator d quarks combine with the u quark to form a neutron, giving $\Sigma^- \rightarrow n + e^- + \bar{\nu}_e$ overall.

The first attempt to understand these types of decay was to assume that the universality of W bosons couplings to leptons could be extended to quarks, so that the u, d and s quarks all coupled to the W boson with the same strength as the lepton couplings. Although this assumption explained many decays involving leptons, others that had been observed were forbidden in this simple scheme. The solution was proposed by the Italian physicist Nicola Cabibbo who invoked the idea of mixing that was used in neutrino oscillation. In the Cabibbo model, the *d* and *s* quarks that participate in the weak interaction are not the same ones that participate in the strong interaction. Instead, they are linear combinations, the mixing being given by a parameter that has to be found from experiment – the *Cabibbo angle*. It was possible to choose this parameter so that previously forbidden decays were allowed to occur; in practice the lifetimes of all hadron decays that result in leptons in the final state could be successfully predicted in terms of just this one parameter. The Cabibbo angle turned out to be small and also provided an explanation for why decays where strangeness is not conserved were typically suppressed compared to those that conserved strangeness.

Hadrons can also decay to other hadrons without leptons appearing in the final state – for example, a lambda to a proton and a pion ($\Lambda \rightarrow p + \pi^-$). These are not described by the Cabibbo scheme but there are rules that govern this type of decay and which are verified by experiment; one of these is that the strangeness quantum number cannot change by more than one unit.

Symmetries

Symmetries, and the conservation laws that follow from them, play a central role in physics, nowhere more so than in particle physics. We have seen that the charged weak interaction does not conserve strangeness, unlike the strong and electromagnetic interactions. Another way of saying this is that the weak interaction displays the least symmetry of the three interactions. Parity and charge conjugation are other examples. (See the text box below for the definitions of parity and change conjugation, and the relation between symmetries and conversation laws.) Parity and charge conjugation are known to be conserved in strong and electromagnetic interactions. Consider, for example, an excited state of a nucleus. When the nucleus decays, either by the strong or electromagnetic interactions, the combined parity of the final nucleus and the emitted particle is the same as that of the excited state: parity is conserved. Transitions that do not conserve parity are not observed to a very high degree of accuracy.

SPACE-TIME SYMMETRIES AND CONSERVATION LAWS

Symmetries and conservation laws are closely related. An example is translational symmetry; the fact that all positions in space are physically indistinguishable. What this means is that when a closed system of particles, that is, one on which no external forces are acting, is moved as a whole from one position in space to another, its physical properties are unchanged, or invariant. This symmetry leads directly to the conservation of linear momentum. Likewise, the conservation of angular momentum is a consequence of rotational symmetry, the fact that all directions in space are physically indistinguishable, so that the physical properties of a closed system of particles are unchanged by a rotation.

In particle physics, other space-time symmetries are important. One is *charge conjugation*, the operation of changing particles to their anti-particles. The invariance of a system under this operation leads to the conservation of a charge conjugation quantum number C in interactions. A single electrically charged particle does not have a value of C but an uncharged particle can have a value of $+1$ or -1, depending on its quantum-mechanical properties. An electrically neutral combination of charged particles can also have a non-zero value of C. For example, a system comprising a positive and a negative pion will still be a positive and negative pion if both particles are replaced by their anti-particles, because in this case charge conjugation only reverses the signs of their charges.

SPACE-TIME SYMMETRIES AND CONSERVATION LAWS (*cont.*)

Another symmetry is *parity*, the operation of reversing the sign of all spatial co-ordinates of a system. A quantity is said to have *odd* (or *even*) parity if it changes (or does not change) sign under this operation and an associated parity quantum number P will be conserved in interactions that are invariant under the parity operation. The elementary particles of the standard model, whether charged or uncharged, also have intrinsic parities: quarks and leptons all have positive parity and anti-quarks and anti-leptons have negative parity. The corresponding quantum numbers are $P = +1$ and $P = -1$ respectively.

Besides the intrinsic values of the C and P quantum numbers, a system of particles has an additional contribution if there are orbital angular momenta between its constituents, unless those momenta are zero. The total values of the quantum numbers P and C are obtained by combining the two contributions, orbital and intrinsic. For the quantum numbers discussed in earlier chapters, such as strangeness, combining simply means adding, but parity and charge conjugation are different. Two or more of them are combined by multiplying the numbers, so that the result is always $+1$ or -1. The parities of hadrons can be deduced from their quark structure by combining the parities of their quark constituents. Thus all three pions have a parity quantum number P of -1, because they are clusters of a quark and an anti-quark with no orbital angular momentum between them.

What happens in weak interactions is, however, entirely different – and was totally unexpected. In the early 1950s, two mesons were observed, one of which decayed to two pions and the other to three pions, with different lifetimes but both consistent with the decays being via the weak interaction. Also, in both cases the final pions had zero orbital angular momentum. Using the fact that the parity of the pion is negative, it followed that the two mesons had different parities. Yet, the mesons had essentially the same mass values, strongly suggesting that they were manifestations of the same particle.

This puzzle prompted two Chinese-American theorists, Tsung-Dao Lee and Yang, to re-examine carefully all known experiments involving weak interactions. After several weeks of intense work, they came to the conclusion that, whereas there were many experiments that established parity conservation in strong and electromagnetic interactions to a high degree of accuracy, no published experiments on the weak interaction had any bearing on the question of parity conservation. This finding, that for many years parity conservation in the weak interaction was asserted without experimental support, was described by Yang as 'startling'. What was more startling was the prospect that parity might be violated. Most physicists doubted that this would turn out to be true. Feynman even made a fifty-dollar bet with a colleague that experiments would confirm parity conservation.

To settle this important question – and Feynman's bet – physicists needed an experiment that would unambiguously test whether parity was conserved in the weak interaction. Lee and Yang had suggested several such experiments in their paper, one of which involved measurements on the beta decay of cobalt-60, a radioactive form of cobalt. The nucleus of cobalt-60 has a spin and hence a magnetic moment. The experiment involved placing a sample of cobalt-60 in a strong magnetic field so that the magnetic moments, and hence the nuclear spins, were aligned in the same direction. Applying the parity operation to this system would reverse the momentum of the electrons but leave the direction of the nuclear spin unchanged. So if parity were conserved, there would be the same number of electrons emitted in the direction of the nuclear spin as were emitted in the opposite direction.

Another way of looking at this is in terms of *mirror symmetry*, where a new system is produced by reflecting the co-ordinates in a hypothetical mirror. (When you look in a mirror, your right side

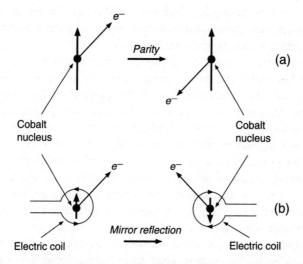

Figure 7.5 Action of (a) parity and (b) mirror reflection on the decay of the cobalt nucleus. The thick lines show the direction of the spin of the colbalt-60 nuclei and the thin lines show the direction of the momentum of the emitted electron

looks like your left side but you are still standing upright.) This is shown in Figure 7.5b, where it is assumed that the magnetic field that aligns the spins of the colbalt nuclei is produced by an electric coil. The right-hand side of this figure is equivalent to reversing the spins of the nuclei (because the magnetic field is reversed), so if mirror symmetry were exact, the direction of emission of the electrons would be independent of the direction of the cobalt-60 spins.

The experiment, simple in concept but difficult in practice, was carried out in 1957 by another Chinese-American physicist, Chien-Shiung Wu. She cooled the cobalt nuclei to the very low temperature of 0.01 K, to reduce their thermal motion. This increased the degree to which the nuclear spins aligned.

Her result was unequivocal: the two distributions were not the same and the electrons emerged preferentially in a direction parallel to the nuclear spin. Parity is not conserved in the weak interaction or equivalently, nature provides an absolute meaning to right and left. Feynman lost his bet.

Parity non-conservation can be demonstrated in many simpler processes, including the decays of the positive and negative charged muons. Because a μ^+ changes to a μ^- by charge conjugation, these decays offered scientists the additional possibility of testing whether charge conjugation is conserved. In this case, the angular distributions of the emitted electrons would be the same for μ^+ and μ^-. The experiments were again unambiguous: as with parity, charge conjugation is violated by the weak interaction.

The discovery of parity and charge conjugation violation was a watershed. The effect is large and essential to understanding the weak interaction. Take, for example, the way this plays out for neutrinos, assumed to be massless. Measurement of the component of a spin-$\frac{1}{2}$ particle in any direction can yield one of two possible values: $+\frac{1}{2}$ or $-\frac{1}{2}$. If the spin direction is that of the momentum of the neutrino (such that a spin measurement in that direction yields a value $+\frac{1}{2}$), the particle is called a *right-handed* neutrino. Similarly, a *left-handed* neutrino is one where the spin projection and momentum are in opposite directions. Remarkably, Wu's experiment implies that only left-handed neutrinos and right-handed anti-neutrinos participate in the weak interaction, which prompted remarks that God was 'a weak left-hander'. This was first verified for the electron neutrino in 1958, in an elegant experiment that directly measured the handedness of the electron neutrino. It revealed that nature really does distinguish between right and left – and 'prefers' the latter.

For particles with mass, the situation is not so simple. However, provided their energy is much greater than their mass, so that their mass can be neglected, the result still holds. Only the interactions of left-handed charged leptons and right-handed anti-leptons occur in this approximation. Consider two decays of pions, $\pi^+ \rightarrow \mu^+ + \nu_\mu$ and $\pi^+ \rightarrow e^+ + \nu_e$: both are charged weak interactions with the same intrinsic coupling strength but the energy released in the latter decay, to a positron and an electron neutrino, is more than four times that of the former, to a muon and a mu neutrino, so the expectation would be that the decay to a positron would be preferred. Experimentally, this is not what happens. Instead, pion decay to a muon proceeds at a rate ten thousand times that of the decay to a positron. Part of the reason for this is that the final positron, but not the muon, is highly relativistic and so in the approximation of neglecting the positron mass, the restriction of handedness applies to the positron and this decay is forbidden. Similar suppression factors occur in other forbidden states.

8

Charm, bottom and top

The second half of the 1950s was a dramatic time for weak interaction physics. The violation of parity, long believed to be a cornerstone of theory (although as Lee and Yang showed, utterly without foundation) stunned the particle physics community. Physicists' intuition had failed completely. However, rising from this shock came a deeper understanding of weak interactions and the vital role played by symmetries. At the start of the 1960s the baton was passed to theorists, who, from symmetry considerations, took the lead in proposing new quarks to solve theoretical problems in both strong and weak interactions. Their imaginative leaps of invention were subsequently vindicated by experiments and were further steps towards the standard model.

Charm quarks

By the dawn of the '60s, two generations of leptons had been observed, the electron and muon and their neutrinos, giving four leptons in total. Yet only three quarks, u, d and s, were known to exist. This was asymmetric and physicists have a strong preference for symmetries. For reasons little other than restoring the balance, several people suggested there should be a fourth quark, to form a second generation with the strange quark. One of these suggestions came in 1964 from two Americans, Sheldon Glashow and James Bjorken, who whimsically named it the *charmed quark* (*c*) and the new quantum number it carried *charm*, denoted *C* (not to be confused with charge conjugation), with a value of +1. However, in the total absence of any evidence for such a particle, their proposal lay dormant for several years.

One scientist who was interested in pushing this idea forward was Glashow himself. He had been working for some time on a theory that would remove the troublesome infinities from calculations in weak interactions. He was by no means the first person to attempt this but he was the first to produce a practical theory of the weak and electromagnetic interactions. Glashow's ideas were developed by another American, Steven Weinberg, and independently by Abdus Salam, a Pakistani physicist working in England. One consequence of their work is that there had to exist a neutral boson analogous to the charged W bosons; the Z^0 boson.

Glashow's theory was successful in describing the interaction of leptons, but he was faced with a problem when it came to quarks. The existence of the Z^0 implied the existence of the *weak neutral interaction*, similar to the weak charged interaction. While the weak charged interaction involved the exchange of W bosons, the neutral version involved the exchange of Z^0 bosons. Moreover, because the d and s quarks have the same charge, they would transform into each other by the exchange of a Z^0. Therefore, Glashow expected to see weak neutral interactions that did not conserve strangeness, for example, the decay of a kaon to a muon pair ($K^0 \rightarrow \mu^+ + \mu^-$). However this decay, and many other strangeness–changing weak neutral interactions, had never been seen.

Undeterred, in 1970 Glashow, in collaboration with the Greek physicist John Iliopoulos, and the Italian Luciano Maiani, realised that if there were a fourth quark that formed a second generation with the strange quark, the additional quark would give rise to new Feynman diagrams whose contributions would exactly cancel the problematic ones that gave rise to the unobserved reactions. This is because the weak interaction that acts on a *ud* quark pair is the same strength as that acting on a *cs* pair but with the opposite sign. This is the GIM (pronounced 'jim') mechanism, named from the initials of its three authors. The GIM mechanism was a bold idea that could have been ridiculed as theorists playing games. There was no evidence that the Z^0 even existed and yet Glashow and his colleagues were using its presumption to justify inventing yet another hypothetical particle. Moreover, the existence of a charm quark would raise an obvious question: where were the particles with a non-zero charm quantum number? If the charmed quark was very light, charmed hadrons, that is, hadrons with at least one *c* quark, would have been seen in existing experiments; if the charmed quark was very heavy, the GIM mechanism would not suppress strangeness-changing weak neutral interactions. There was a fairly narrow range of masses available where the scheme was viable. Fortunately, this range was becoming accessible in the accelerators beginning to be used at that time.

The cleanest reaction that can be used to search for evidence of charm is electron-positron annihilation. Using a collider at Stanford University, in November 1974 a team led by the American Burton Richter announced they had found evidence in the form of a meson. The meson was a bound state of a charmed quark and its anti-particle, with a total charm quantum number of 0. The state was referred to as having *hidden charm*. At the conference where Richter presented these results, another American physicist, Samuel Ting, from Brookhaven National Laboratory, also presented evidence for charm. His team was searching for new states by studying the electron-positron pairs found amongst the debris produced in high-energy proton collisions with nuclei. Ting's approach borrowed from the technique used to identify hadron resonances. In a sense, this was the inverse of the process being studied by Richter. Both groups had accumulated data that showed a huge enhancement in the total effective mass of these electron–positron pairs at a value of 3097 MeV/c^2. Ting called the particle J; Richter called it *psi* (ψ). As a result, physicists refer to it by the rather clumsy notation of J/ψ. It is also known as *charmonium*, a reference to the bound state of an electron and a positron, known in atomic physics as positronium.

Charmonium allowed physicists to deduce some properties of the c quark. From the mass of the J/ψ, it was calculated that the c quark has a mass of about 1500 MeV/c^2, far heavier than the other quarks, whose constituent masses had been estimated to be about 350 MeV/c^2 for the u and d quarks and 500 MeV/c^2 for the s quark. Within a year many other $c\bar{c}$ bound states had been discovered at several laboratories. Figure 8.1 shows the measured ratio R for the production of hadrons to the production of muon pairs ($\mu^+\mu^-$) in electron-positron annihilations. The extremely narrow peaks at 3097 MeV/c^2 and 3886 MeV/c^2 are indicated by arrows, because they extend far above the scale depicted in the figure.

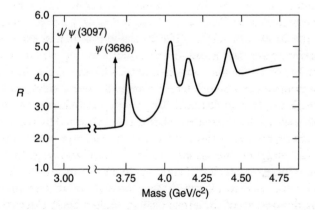

Figure 8.1 Charmonium production in electron-positron annihilations

How could experimentalists be so sure that the charmonium states really were made of a c quark and a $\bar{c}$ quark and not a pair of lighter quarks? The answer is to be found in their decay properties. If charmonium states do indeed have the structure $c\bar{c}$, their decays would be expected to occur via the strong interaction, resulting in a pair of mesons, each of which would contain a charmed quark (or anti-quark), so charmonium states would have decay widths measured in MeV. However, the lower-mass states had widths that were much smaller than this; measured in keV. This would only occur if the lowest mass meson with a single charmed quark was such that a pair of them had a mass greater than the J/ψ; energy conservation would forbid this particular mode of decay and more complicated decay routes would have to be followed. Above the threshold for producing a pair of charmed mesons, the widths of charmonium states would be measured in MeV, as observed. In 1976, two years after Ting and Richer announced the existence of hidden charm, the first meson with a single charmed quark – exhibiting so-called *naked charm* – was found. The mass of this meson was about 1870 MeV/c^2, confirming this explanation. Later experiments produced evidence for resonances of charmed mesons and baryons with naked charm.

The discovery of the J/ψ is often called the *November Revolution*; not only was it a breakthrough in understanding the weak interaction but it also helped QCD gain more general acceptance. For example, QCD provides an explanation for why the more complicated decay routes for charmonium states with masses below the threshold for producing pairs of charmed mesons were suppressed. A typical decay is to three pions; in QCD, both the decaying particle and the three pions are colourless, so can only be connected by a combination of gluons that is also colourless, not by the exchange of a single gluon. The smallest number of gluons that satisfies this requirement and also has the correct charge conjugation quantum number (also conserved in the decay) is three. This is a higher-order process and hence these decays are suppressed.

QCD also provides an explanation of the relative lifetime of the J/ψ to another spin-1 meson, phi (ϕ), which is a member of the spin-1 resonance multiplet that includes the rho meson. The phi meson is a bound state of an s quark and its anti-particle, $\phi = s\bar{s}$, so is similar to the J/ψ. QCD's property of asymptotic freedom predicts that at high energies, or equivalently short distances, the strong force decreases in strength. Thus at the high energies of the J/ψ, the c quark and its anti-particle are less strongly attracted to each other than an s quark and its anti-particle at the lower energy of the phi. This means the probability of them annihilating is reduced, leading to the longer lifetime observed for the J/ψ, almost one hundred times longer than the phi.

The third generation

After the discovery of the tauon in 1975, theorists suggested that there had to be a third generation of quarks to restore the symmetry between quarks and leptons. The quark numbers for this new generation were originally labelled *truth* for the quark with charge $\frac{2}{3}$ and *beauty* for the quark with charge $-\frac{1}{3}$ but after a while this was considered a little too exotic even for particle physicists and the more mundane names *top* and *bottom* were coined. These names refer to their positions if each is arranged as a pair of quarks, with the quark having the larger charge above the one with smaller charge. The bottom quark b carries a new quantum number, *bottom B* with a value -1, and the top quark t carries a quantum number *top T* with a value $+1$. Both of these quantum numbers are conserved in the strong interaction, as with strangeness and charm.

It took a couple of years before the first b quarks were found. In the summer of 1977, experimenters looking at the interaction of high-energy protons with nuclei identified a particle with a mass of 9.45 GeV/c^2; about ten times the mass of the proton. They called it the upsilon (Υ). Its existence was confirmed a year later when the first electron–positron collider with sufficient energy to produce the upsilon particle became operational. By observing its decays, physicists determined that the upsilon is a bound state of a b quark and its anti-quark, $b\bar{b}$. Its mass implies that the mass of the b quark is about 4.5 GeV/c^2. The upsilon was the first of a series of $b\bar{b}$ states, named *bottomium*. As in the case of charmonium, below a particular mass their decay widths are measured in keV, while above this mass they are measured in MeV. The critical mass is where decays to a pair of mesons, each with a b or $\bar{b}$ quark are possible. The lightest such state is the B meson with a mass of 5.3 GeV/c^2. Later experiments found resonance mesons with *naked bottom* (perhaps beauty would have been a better choice after all), as well as baryons containing a b quark. Just as charmonium provided evidence that supported QCD, so did bottomium, because the masses of the $c\bar{c}$ and $b\bar{b}$ states, relative to their ground states, show very similar patterns, confirmed by detailed calculations, indicating that that the strong force between quarks is flavour-independent.

Finding the top quark took much longer. It was finally found in 1995, using the CDF detector at Fermilab's Tevatron collider. Each of the colliding beams at the Tevatron has a maximum energy of 1 TeV. At such high energies, pairs of top quarks and anti-quarks can be produced. Because the mass of the top quark is greater than that of the W boson, the top quark can decay to a real $W(t \rightarrow q + W^+)$, where to conserve charge, q could be a d, s or b quark. In practice, the only significant decay is to a bottom quark $(t \rightarrow b + W^+)$. The bottom quark produces a jet of hadrons, via hadronisation, and the W boson decays, predominantly to either pairs of light quarks and anti-quarks or to leptons. In the former case, further jets of hadrons are produced. The $\bar{t}$ quark decays in a similar way, so that overall, four jets of hadrons are

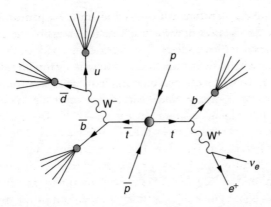

Figure 8.2 Schematic diagram of the production and decay of a $t\bar{t}$ pair in $p\bar{p}$ annihilations

observed (two due to the bottom quarks and anti-quarks and two due to the light quarks from $W^{\pm}$ decays), plus a charged lepton or anti-lepton (either a muon or an electron) and an undetected neutrino or anti-neutrino, as appropriate. A typical 4-jet event is shown schematically in Figure 8.2. (The reconstruction of a real $t\bar{t}$ event was shown in Figure 4.3. In it one can see the four jets and a positron that were produced.) The mass of the top quark can be found from the energies of the particles and yields a value of about 180 GeV/c^2.

This enormous mass, approximately two hundred times the mass of the proton – about the mass of an atom of gold – has some immediate consequences, one of which is on the decay rate of the top quark. Weak decay rates increase as the energy released during the transition increases and decrease with decreasing energy released (recall the long lifetime of the neutron). The increase is very rapid and a major contribution to the fact that the top quark has a lifetime of approximately 10^{-25} s, much shorter than would be expected even for a decay via the strong interaction.

Because of its fleeting existence, there are few opportunities to study the properties of the top quark: for example, there is too short a time between the production of a top quark and its decay for top quarks to form into clusters due to the strong interaction. Thus, unlike for charm and bottom, there are no hadrons with a non-zero value of top. Neither is there a 'toponium' equivalent of charmonium and bottomium.

Quark spectroscopy

No one really knows if the discovery of the top quark marks the end of the story or if more generations of quarks remain to be discovered. There is strong evidence that there are no more lepton generations with masses below the mass of the W boson, so on the grounds of symmetry, physicists do not expect to find any more quark generations unless they have truly enormous masses, in which case their lifetimes would be immeasurably short.

What are the consequences of the extra quarks we do know about? The existence of charm and bottom quarks leads to a much richer hadron spectroscopy, because there are far more combinations of internal quantum numbers to be considered. The simple diagrams of the Eightfold Way now require more dimensions and are usually displayed in sections, where each layer corresponds to a fixed value of one internal quantum number. One example is shown in Figure 8.3, which illustrates the case of the spin-$\frac{3}{2}$ baryons made from any combinations of the u, d, s and c quarks, where each layer has a fixed value of charm. Not all the predicted states of the extended quark model have been discovered, as some are very difficult to access experimentally, but the gaps are gradually being filled. The latest to be identified, the *omega-minus-bottom* baryon (Ω_b^-) was found in late 2009. This state is like the famous omega-minus Ω^- that provided important evidence for the original quark model but one of the s quarks has

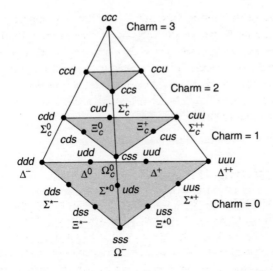

Figure 8.3 The spin-$\frac{3}{2}$ baryons made from *u*, *d*, *s*, and *c* quarks

been replaced by a *b* quark, so that it is a cluster of two strange quarks and one bottom quark.

Now we have met all three generations of quarks, we can systematically examine their weak decays. This is a good time to re-emphasise that by decay, particle physicists mean the decay of a quark bound in a hadron, so that spectator quarks are present, although this does not apply to the top quark. In the first generation, transformations between the *u* and *d* quarks are linked by the emission of either a W^+ or a W^- boson. The same is true for transformations between the quarks in the second generation, the *c* and *s* quarks; moreover, the strengths of these transformations are equal. The third generation of quarks seems to share this property. There also appears to be near equality in the *W* boson couplings to these quarks, compared with its couplings to leptons. The small discrepancy observed

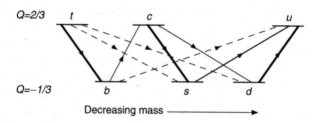

$Q=2/3$ t c u

$Q=-1/3$ b s d

Decreasing mass ⟶

Figure 8.4 Systematics of the weak decays of quarks

can be accommodated by the Cabibbo mixing scheme. Unlike leptons, where lepton number conservation forbids transitions between generations, quarks make cross-generational transitions, for example between c and d quarks and between s and u quarks. While these transitions are possible, some are favoured over others. Similar suppressions occur when mixing takes place between all three generations.

The actual pattern of decays observed is shown in Figure 8.4. Upward-pointing arrows represent transitions where an electron-neutrino pair, $e^- \bar{\nu}_e$, is emitted and downward-pointing arrows represent transitions where a positron-neutrino pair, $e^+ \nu_e$, is emitted. The solid lines are the dominant transitions, with their relative strengths being denoted by thick or thin lines, the thicker lines indicating the more preferred transitions. The dotted lines show transitions that also exist but are suppressed with respect to the others.

CP violation and the standard model

The weak interaction violates both parity and charge conjugation separately. However, the combined action of parity and charge conjugation, *CP*, was found to be conserved. Then in 1964, an experiment by the American physicists James Cronin and Val Fitch produced surprising evidence for CP violation where neutral kaons decayed to pions. This violation of CP implied that another symmetry, *time-reversal*, T, was being violated. Time reversal is the operation where the sign of the time co-ordinate in all equations is reversed. Although this does not lead to a quantum number in the way parity and charge conjugation do, if the equations remain the same after this operation, time reversal is said to be conserved. When CP is violated it implies T violation, because any field theory that obeys relativity and quantum theory has to remain unchanged under the combined action of CPT – a general result called the CPT *theorem*. The violation of CP has been confirmed in other decay modes of neutral kaons as well as the decays of neutral B mesons, the lightest meson containing a bottom quark. The circumstance under which this occurs is unusual and offers a unique window into the workings of the standard model.

There are two neutral kaons: the K^0 and its anti-particle $\overline{K}^0$. They both have spin 0, with strangeness $S = +1$ and $S = -1$, respectively. It is these states that participate in the strong interaction. However, they do not directly participate in the weak interaction; instead two combinations do – mixing at work. These combinations are called K^0-*short* (K_S^0) and K^0-*long* (K_L^0), because their lifetimes are not the same; the lifetime of K_S^0 is approximately 10^{-10} s, whereas the lifetime of K_L^0 is 5×10^{-8} s. Their masses are essentially identical but their decay modes differ. In particular, when they decay to pions, K_S^0 decays to two pions, whereas K_L^0 decays to three pions.

This is a very significant difference; although the K^0 and $\overline{K}^0$ do not have definite values of CP, combinations can be constructed that do. These combinations are called K_1^0 and K_2^0 and are assigned CP values of $+1$ and -1, respectively. A state of two pions with a total angular momentum of zero (the spin of the kaon) has a CP value of $+1$ and a state of three pions, also with a total angular momentum of zero, has a CP value of -1, so if CP

were conserved in the weak interactions, it would be natural to match the observed states K_S^0 and K_L^0 with the theoretical states K_1^0 and K_2^0. This identification usually gives a good account of the weak decays of neutral kaons but the Cronin and Fitch experiment found that in about one event in a thousand, the K_L^0 decayed to two pions, violating CP conservation. This held true in later experiments that examined other decay modes of the K_L^0.

A particle cannot usually convert to its anti-particle; normally, some conserved quantum number absolutely forbids such a transition. In the case of proton-anti-proton pairs, for example, this is baryon number. For K^0 and $\overline{K}^0$ pairs, there is no absolutely conserved quantum number, so K^0 and $\overline{K}^0$ can convert into each other and CP can be violated. However, because strangeness can only change by one unit at a time in weak interactions, this transition can only happen if two W-bosons are exchanged, which is a higher-order weak interaction and hence suppressed. There are only two other transitions that can occur: the neutral B mesons and the neutral D mesons, the latter being the lightest meson with a charmed quark. These transitions can occur because, like strangeness, the charm and bottom quantum numbers are not absolutely conserved in the weak interaction.

The lightest B^0 meson has a mass of 5.3 GeV/c^2 and a quark structure $\overline{d}b$. The B^0 meson can mix with its anti-particle $\overline{B}^0$ using the same rules used to describe mixing between neutral kaons. The physical particles that take part in the weak interaction are called B_L^0 and B_H^0, where L and H stand for 'light' and 'heavy', although this is a little misleading as the two states have almost identical masses. Unlike their kaon counterparts, which are distinguished by their lifetimes, the B_L^0 and B_H^0 have almost identical lifetimes of approximately 1.5×10^{-12} s; very short compared to the neutral kaons. Because of their very brief lifetimes, it is not possible to form well-defined beams of neutral

B mesons; other means had to be found to study their decay modes, which because of their large masses are numerous.

To overcome this problem, physicists have constructed B-factories, a special type of collider that exploits the properties of a particular upsilon state with a mass of 10.58 GeV/c^2. This is heavy enough to allow it to decay to a pair of the lightest mesons with non-zero bottom quantum number, but not heavy enough to decay to any other final states by the same mechanism. It therefore decays almost entirely to B^+B^- and $B^0\overline{B}^0$ pairs. Tuning the beam energies so that their total energy is equivalent to a mass of 10.58 GeV/c^2 therefore results in a copious supply of B mesons.

Two B-factories have been built. One is in the Japanese KEK Laboratory and the other at SLAC in California. In the KEK machine, a 3.5 GeV positron beam is collided with an 8 GeV electron beam. The total energy created corresponds to the upsilon mass and the asymmetric beam energies ensure that the B mesons are produced with enough momentum to travel a measurable distance before decaying. This is essential, because studies of CP violation often require the measurement of the short time between the production and decay of the B mesons. Experimentalists have achieved this by using a silicon vertex detector just outside the beam pipe in a dedicated detector, Belle, which has a structure typical of the multi-component detectors used at conventional colliders.

In experiments involving neutral kaons, measurements are taken after leaving sufficient time for the short-lived component K_S^0 to decay. At that point only the K_L^0 remains. However, because B_L^0 and B_H^0 have very similar lifetimes, it is not possible to use the same method to observe samples of just one type of particle, so the analysis is more complicated. Nevertheless, the results are clear: CP violation is observed in a number of different decays of neutral B mesons and the size of the effect is far greater than observed in the neutral kaon system. CP-violating effects are also expected for neutral D mesons but these have not been studied to any extent; so far no such effects have been observed.

What do these results imply for the standard model? In mixing between the d and s quarks, a single parameter – the Cabibbo angle – determined the relative fractions of each type that participated in the weak interaction. With a third generation of quarks the situation is more complex, because to describe the mixing between all three quarks, d, s and b, requires four parameters. Three of these are mixing angles, analogous to the Cabibbo angle.

Interestingly, a non-zero value for the remaining parameter implies violation of T invariance and hence, assuming CPT invariance holds, violation of CP conservation. This cannot occur for mixing with only two generations. The magnitude of the effect depends on all the parameters, which can be obtained from a range of decay data. Using these, the prediction is that CP-violating effects will occur in the neutral meson systems and will be largest for the neutral B mesons and much smaller for the neutral K and D mesons. This is entirely consistent with experimental observations, although quantitative predictions are difficult to obtain.

Finally, it is worth noting that both T invariance and CPT invariance can be tested by direct measurements on neutral kaon decays. T-violation may be studied by measuring the time dependence of the difference in the rate for the transition of an initial K^0 to a $\overline{K}^0$ and the corresponding rate for the time-reversed process in which an initial $\overline{K}^0$ makes a transition to a K^0, where the type of kaons are identified by their decay products. The CPLEAR experiment, carried out at CERN in 1998, found this difference to be non-zero; moreover, the data were consistent with measurements of CP violation if CPT invariance was assumed. The same experiment also tested CPT invariance by measuring the difference in rates where the initial particle does not change its identity. In this case the difference was found to be consistent with zero, verifying the *CPT* theorem. CPLEAR closed any potential loopholes in the analysis of *B* meson decays. The success of the mixing model in accounting for all *CP*-violating effects in terms of a single parameter is yet another triumph of the standard model and shows that it is capable of describing even fine details of the weak interaction.

9

Electroweak unification and the origin of mass

The hypothesis of the existence of a neutral force carrier, the Z^0 boson, focused physicists' attention first on observing reactions that were predicted to occur through its exchange, and later on constructing experiments that would reveal both the W and Z^0 bosons as real particles. These efforts provided an understanding of the number of generations of leptons and by implication, the number of generations of quarks. Theorists were also very active, building on the similarities between the weak interactions of leptons and quarks to construct a theory that unified the weak and electromagnetic interactions into a single *electroweak* interaction. Although it is very successful in explaining the data, the electroweak theory contained a potentially serious problem: it initially predicted that all elementary particles have zero mass.

Weak neutral interactions and strangeness

Attempts to deal with the divergences that occurred in the higher-order diagrams of the weak interactions, for example those that involved the exchange of more than one W boson, led theorists to postulate the existence of a neutral particle, the Z^0 boson, the exchange of which led to the prediction of weak neutral interactions. These exchanges are governed by a coupling constant α_Z, similar to the weak charge coupling constant α_W. Both are 'running' coupling constants, that is, they are energy dependent, like those in QED and QCD.

By the introduction of a fourth quark, with a new charm quantum number, the GIM mechanism explained the absence of weak neutral interactions that did not conserve strangeness. However, the question remained whether weak neutral interactions that did conserve strangeness existed. A reaction that could shed light on this involves mu neutrinos, $\nu_\mu + N \rightarrow \nu_\mu + X$, where N is a nucleon and X is a collection of hadrons that conserve all the relevant quantum numbers. For example, in a bubble chamber, an event of this type would display no incoming track, merely a number of outgoing tracks originating from a common point.

The first evidence for such events came in 1973, from a detailed examination of numerous photographs taken in the Gargamelle bubble chamber, which, filled with a heavy liquid, had been exposed to a beam of mu neutrinos at CERN. Figure 9.1 shows a typical interaction, giving a sense of just how difficult it was to identify these events. The incident neutrino enters the chamber from the left but produces no track; it interacts with a neutron in a nucleus of the chamber's liquid, producing a negative pion and a proton, which can be seen clearly. Four photons are also detected through their conversion to electron-positron pairs $e^+ e^-$. The photons almost certainly arose from the decays of neutral pions produced in an initial neutrino–proton interaction and in the subsequent interaction of the negative pions with another atomic nucleus. Weak neutral interactions conserve strangeness, as well as the other quantum numbers, charm and bottom.

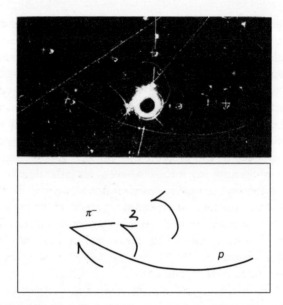

Figure 9.1 Weak neutral reaction in a heavy-liquid bubble chamber (reproduced by permission of CERN)

Electroweak unification and the boson masses

In the theory constructed by Glashow, and developed independently by Weinberg and Salam, that introduced the Z^0 boson into the weak interaction, it was necessary to include Feynman diagrams with exchanged photons. Although these diagrams, considered individually, gave rise to infinities, they cancelled the infinities that arose from the exchange of W and Z^0 bosons. As a result, the weak and electromagnetic interactions are married into a single interaction – the electroweak interaction – and the cancellation happens order-by-order in perturbation theory. This is not accidental but relies on gauge invariance.

The proof of the cancellation, made by 't Hooft, is mathematically formidable and relies on a fundamental relationship between three dimensionless couplings: first, the electromagnetic coupling constant α; second, the weak charged coupling α_W due to W boson exchange; and third, the weak neutral coupling α_Z due to the exchange of the Z_0 boson. This relationship, the *unification condition*, is given in terms of a new parameter, the *weak mixing angle* (also called the Weinberg angle), written θ_W, which specifies the ratio of the W and Z^0 boson masses. The weak mixing angle can be found from experiment by comparing the rates of weak charged and weak neutral interactions at low energies. The value obtained can then be used to predict the masses of the W and Z^0, using the known value of the electromagnetic coupling. When this was first done, the predicted mass values were 78 GeV/c^2 for the W boson and 89 GeV/c^2 for the Z^0 boson. Do these numbers agree with experiment?

At first, there was no way of answering this question: no existing accelerator had high enough energy to produce such heavy particles. In the late 1970s, a momentous decision was taken to build a new system of accelerators at CERN, with the express purpose of looking for the W and Z^0 bosons. In 1976, CERN had commissioned the Super Proton Synchrotron (SPS) – shown in Figure 3.4 – which had as its injector a smaller proton synchrotron, which produced anti-protons. However, only about one anti-proton was produced by every million protons, so it took approximately one whole day to accumulate a usable number of anti-protons. These anti-protons were 'squeezed' to create an intense parallel, and mono-energetic beam, in an ingenious new machine, the anti-proton accumulator (AA), which used van der Meer's cooling technique, described in chapter 4. The AA enabled more than 10^{11} anti-protons to be accumulated from initial batches of about 10^6 anti-protons. Finally, the beam was injected into the SPS and both it and the proton beam were accelerated (circulating in opposite directions around the ring), generating a proton–anti-proton collider.

Two experiments were constructed in an underground area at CERN, one of which was led by the Italian physicist Carlo Rubbia who had been the main driving force behind the decision to build the collider. The experiments used beams with equal energies of 270 GeV, giving a total available energy of 540 GeV. At first sight this might appear to be overkill, because the energy appears to be more than enough to study the proposed reactions ($p + \bar{p} \rightarrow W^+ + X^-$, $W^- + X^+$, and $Z^0 + X^0$). However, the interactions involved a single quark in the proton and a single anti-quark in the anti-proton and these on average will only have about one third of the energies of the proton and anti-proton. So 540 GeV was only just above the threshold energy to produce the W and Z^0 bosons.

The dominant decays of the $W^\pm$ and Z^0 are to pairs of quarks and anti-quarks, which lead to jets of hadrons by hadronisation, but because the spectator quarks also produce jets, it is difficult to disentangle the two sources. But Z^0 bosons also decay to pairs of leptons of any flavour and so the experimentalists used these decay modes in their search. They plotted the total mass of pairs of electrons and positrons and looked for enhancements (the same technique as had been used to find hadron resonances). However, the $W^\pm$ bosons decay to a charged lepton and an undetected neutrino, which posed a greater challenge. The main problem facing the experimentalists was that for every real event there were 10^7 events where hadrons alone were produced, without an accompanying boson!

In 1983, CERN announced clear evidence for the W and Z^0. Later experiments confirmed CERN's results. One of the experiments used the CDF at Fermilab and produced the plots of the total mass of charged lepton pairs shown in Figure 9.2. The measured W and Z^0 masses matched the theoretically predicted values. These data conferred great confidence in the electroweak theory as a critical component of the standard model.

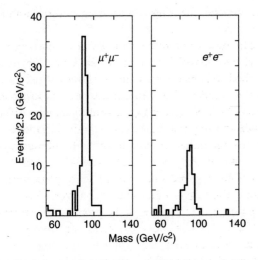

Figure 9.2 Z^0 peaks observed in e^+e^- and $\mu^+\mu^-$ mass distributions

The discovery of W and Z^0 with the predicted masses opened up a host of new puzzles. Could there be further generations? Why do the forces have their observed strengths? CERN rose to the challenge of answering these questions by building another new accelerator, the Large Electron Positron Collider (LEP), which began operation in 1989. Until its closure in 1996, LEP produced beams with energies of 45 GeV, enough to produce copious numbers of Z^0 bosons. The beam energies needed were less than those in the SPS collider because the reactions studied were between leptons, not quarks. LEP was essentially a Z^0 factory, producing millions of these particles.

The LEP experiments provided a very accurate measurement of the Z^0 mass. When this more accurate value became available in the 1990s, it forced physicists to update the predictions based on the electroweak theory. Foremost, the data led to a better determination of the weak mixing angle and to a revised prediction of the Z^0 mass that matched its newly-measured value.

The original prediction had neglected higher-order effects in the weak interaction, including some that involved virtual top quarks. Once these were taken into account, agreement was restored, lending further support to the fine details of the electroweak theory. Because these very small corrections involved the mass of the top quark, it was now possible to make an estimate of that mass. By 1994, LEP experiments were able to predict that the most probable value was in the range 170 to 180 GeV/c^2; very close to the experimental value found when the top quark was eventually discovered.

How many neutrinos?

Earlier, I posed the question of whether we should expect to find further generations of quarks. Although this remains open, there is some evidence from the decay of the Z^0 particle that indicates no more lepton generations exist. Because it is so heavy, the Z^0 can decay to many final states: hadrons and pairs of leptons of all flavours, both charged and uncharged. Each of these possible final states can be measured experimentally, with the exception of the decay to a neutrino and its anti-particle. All flavours of neutrinos and anti-neutrinos are possible, but because of lepton universality we know that the probability for decay to each flavour is equal. Thus, if there were N generations of neutrinos, their total contribution to the decay width of the Z^0 would be N times the value for the width for decay to any pair, for example to neutrino pairs such as $\nu_e \bar{\nu}_e$. This contribution can be calculated from the standard model. By gathering all these data, N can be determined. The results are shown in Figure 9.3, as curves for predictions of the production cross-section of hadrons in the vicinity of the Z^0 mass, for three different assumed values of the numbers of neutrino generations. The experimental data, from an experiment at CERN, are shown as black circles. The data

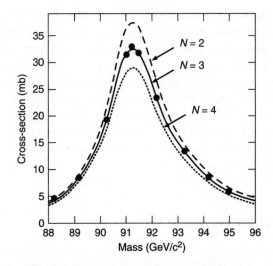

Figure 9.3 Production of hadrons in e^+e^- annihilations in the vicinity of the Z^0 mass for different numbers of neutrino generations

clearly favour three generations of particles and rule out other values.

There is, however, a qualification for three generations: the argument assumes that neutrinos all have masses that are less than half the mass of the Z^0 boson, which would mean that decays to two neutrinos would not be forbidden by energy conservation. For this reason, the result only applies to light neutrinos and the possibility of further neutrino generations cannot completely be ruled out.

Electroweak reactions

In the electroweak theory, any process in which a photon is exchanged also allows a Z^0 particle to be exchanged. At energies

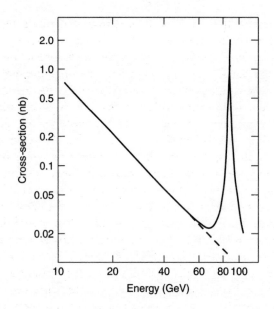

Figure 9.4 Cross-section for the reaction $e^+ + e^- \rightarrow \mu^+ + \mu^-$

much less than the mass of the Z^0, this is not apparent; electromagnetic and weak interactions remain distinct, with the former dominating the latter. As the energy increases, so does the importance of Z^0. Figure 9.4 illustrates the simple reaction involving electron–positron collisions producing muons ($e^+ + e^- \rightarrow \mu^+ + \mu^-$) over a range of energies. At low energies, the cross-section is due mostly to the electromagnetic interaction and decreases with increasing energy. However, as the energy increases, the Z^0 contribution, which increases with energy, becomes more dominant. This continues until the mass of the Z^0, about 90 GeV/c^2, is reached, at which point real Z^0 particles are produced. Beyond this energy, photon and Z^0 exchanges are comparable.

Another consequence of the electroweak theory is that reactions that would normally be viewed as electromagnetic should exhibit parity violation because of the weak interaction effects resulting from Z^0-exchange contributions; or equivalently, they should distinguish left from right. Several experiments have demonstrated this to be true. For example, in elastic scattering of electrons from protons, where the electrons are prepared with definite spin orientation, it has been found that left-handed electrons, that is, those in which the electron's spin is oriented in the opposite direction to its momentum, interact more readily than right-handed electrons, where the spin is oriented in the same direction as the momentum. Moreover, the size of the very small difference in the cross-sections of electrons with different handedness is successfully explained in the electroweak theory. Another experiment, conducted in 2004, involved scattering electrons prepared with a definite handedness from the atomic electrons in hydrogen. The difference in the cross-sections for the right-handed and left-handed electrons was measured and although extremely small, about one part in 10^7, it matched what was expected from the standard model. As in weak charged interactions, parity is unambiguously violated in weak neutral interactions.

The origin of mass

The successes of the standard model are very impressive but they are not unequivocal. The model's component field theories – QCD and the electroweak theory – are gauge theories, which means they have the property of gauge invariance. In its simplest form, gauge invariance requires that the spin-1 gauge bosons – the force carriers – have zero masses if they are the only bosons present. This is acceptable for QED and QCD, because the gauge bosons are photons and gluons, which do indeed have zero masses, but for the weak interaction, the force carriers are the heavy W and Z bosons. Gauge invariance also plays a crucial role in the unified electroweak theory, where it is needed to ensure the cancellation of the divergences that occur in individual higher-order Feynman diagrams, that is, to ensure that the theory is renormalisable. In this case, gauge invariance requires that all of the fundamental particles – quarks, leptons and bosons – have zero masses if gauge bosons are the only bosons in the theory. These predictions are obviously in contradiction with experimental evidence; a conundrum known as the *origin of mass*.

A possible solution to this problem came from the work of the British physicist Peter Higgs, and others, who had considered the consequences of assuming the existence of a new field, now called the *Higgs field*, with unusual properties in its behaviour in a state without particles, the *vacuum state*. Other fields, such as those due to electromagnetism, are assumed to have a zero value in a vacuum state, which agrees with common sense. However, the Higgs field is assumed to be non-zero. Gauge invariance no longer holds in this state and so particles do not have to be massless. Thus, gauge bosons can acquire mass without violating the gauge invariance of the interaction or destroying renormalisation. This is the *Higgs mechanism*.

This form of symmetry-breaking, where the gauge invariance of the interaction remains intact, is called *spontaneous symmetry breaking* and is regularly found in other branches of physics. An oft-quoted example is that of an iron magnet, the magnetism of which is the cumulative effect of all the individual magnets that result from the spins of the electrons within the iron's atoms. The forces that couple the spins are unchanged if the magnet is rotated, and at high temperatures the net magnetisation becomes zero. Yet, when the temperature falls below a critical value, the spins align in a particular direction, resulting in a net magnetisation, breaking the rotational symmetry. Although the magnetisation could point in any direction, the other properties of the magnet do not depend on the actual direction, so the rotational invariance of the interaction remains unchanged.

The Higgs mechanism received scant interest when it was first published in 1964. It was viewed as a mathematical curiosity that introduced a new field, for which there was no experimental evidence, to solve a theoretical problem. Only later did Weinberg recognise that this mechanism was exactly what was required to unify the electromagnetic and weak interactions in a way that was gauge invariant but did not lead to zero masses for the fundamental particles.

The Higgs mechanism has three main consequences for the electroweak theory. First, the $W^{\pm}$ and Z^0 bosons acquire mass in a ratio determined by the Weinberg angle but the photon remains massless. Second, in the same way that photons are the particles associated with the electromagnetic field, particles – the eagerly anticipated Higgs bosons – associated with the Higgs field must exist. In the basic implementation of the Higgs mechanism, the prediction is that there exists a single spin-0, electrically neutral Higgs boson, H^0. (This is the simplest assumption. Some extensions of the standard model predict the existence of more than one Higgs boson, not all of which would be electrically neutral.) Third, fermions, such as leptons and quarks, can also acquire mass through interactions with the Higgs field and the Higgs boson. In the years since its invention, the Higgs mechanism and the associated Higgs boson have become central to the standard model.

Finding the Higgs boson

The existence of the Higgs boson is the most fundamental prediction of the standard model yet to be verified by experimental observation. One difficulty is that the mass of the Higgs boson is not predicted by the model. A tentative estimate comes from predictions of the mass of the Z^0: just as it was necessary to take account of higher-order diagrams involving top quarks, diagrams with virtual Higgs bosons are also indispensible. These lead to the estimate that the mass of the Higgs boson is less than about 190 GeV/c^2; some theorists think this could even be refined to a somewhat lower value. However, until the search for the Higgs boson is fulfilled, the actual value of the mass remains unknown.

Unlike its mass, the couplings of the Higgs boson to other particles are predicted. The interactions of the Higgs boson with pairs of fermions (quarks or leptons) are characterised by a dimensionless coupling constant, α_H, which is essentially proportional to the masses of the particles to which it couples. The Higgs boson is therefore thought to couple very weakly to light particles such as neutrinos, electrons, muons and u, d and s quarks, and much more strongly to heavy particles such as $W^\pm$ and Z^0 bosons and, presumably, b and t quarks. Thus the attempts to observe the Higgs boson have been made more difficult by the need to first produce the very heavy particles to which it couples. The boson is predicted to have a rich spectrum of decay modes, the relative probabilities of which depend strongly on its mass.

Although scientists have so far failed to find the Higgs boson, the experiments do provide a limit on the value of its mass. The best results have come from the LEP collider, where experimenters studied the reaction $e^+ + e^- \rightarrow H^0 + Z^0$, concentrating on possible decays of the Higgs to $b\bar{b}$ pairs, because these would have the largest couplings to the boson. In practice, the quarks would be observed as jets containing short-lived hadrons with a non-zero bottom quantum number. The results were tantalising. By the time LEP closed in November 2000, to make way for another experimental project, it had demonstrated that no Higgs bosons existed with a mass less than 113.5 GeV/c^2. Some evidence had been obtained for the existence of a Higgs boson with a mass of 115 GeV/c^2, very close to the upper limit of masses accessible to LEP, but unfortunately there were insufficient data for the result to be statistically convincing. The search for the Higgs boson has also been carried out at the Fermilab's Tevatron. Searches there have concentrated on the production of a Higgs boson in association with either a $W^{\pm}$ or a Z^0 boson. If the mass of the Higgs boson is less than approximately 130 GeV/c^2, then, as at LEP, the most likely decay is $H^0 \rightarrow b\bar{b}$. However, no definite signal for the Higgs boson has been observed.

If the Higgs boson exists, it will most probably be seen at the LHC, where two of the experiments, ATLAS and the Compact Muon Solenoid (CMS), have been optimised to search for the particle with a mass up to 1 TeV/c^2. This number is not a random choice: to obtain finite results from calculations in the electroweak theory, it was necessary to include both photon and Z^0 exchanges. Where the produced particles are real spin-1 W bosons, nonsensical predictions can still result at very high energies. This issue disappears if further diagrams with the exchange of Higgs bosons are included. The energy at which the extra contributions of these bosons become essential can be calculated, because it depends on the strength of the charged weak interaction. It is about 1 TeV, equivalent to an effective temperature of about 10^{17} K, which existed for a mere 10^{-14} seconds after the Big Bang. If the theory is correct, collisions at these energies will create the sort of conditions where Higgs bosons will emerge and the full symmetry of the early universe will be revealed.

At the LHC, experimenters will be paying most attention to the reaction $p + p \rightarrow H^0 + X$. Several different forms are possible for the particles X in the final state, each of which will result in a different signature for identifying the Higgs boson. Which will be the most effective depends to some extent on the actual mass of the Higgs. At present the jury is out, but physicists are optimistic that the particle will be found at the LHC.

If the Higgs boson is found, experimenters will turn to determining its mass and other properties precisely. If it is not found, theorists will be forced to re-think the whole basis of electroweak unification and the origin of mass. No matter how the hunt for the Higgs turns out, interesting times lie ahead.

10

Beyond the standard model

Particle physics is still a relatively young branch of physics, even if you date its birth to the discovery of the electron in 1896. Yet it has a comprehensive theory, the standard model, the success of which is already spurring physicists to construct larger unifying theories to incorporate the strong interaction – even gravity in some cases – to understand why particle masses and forces have the values they do, why there are three generations of elementary particles and why gauge invariance holds such a special place in nature.

Grand unified theories

One area where particle physicists are extending the standard model involves attempts to include the strong interaction in a unification scheme with the electroweak interaction. These are the *grand unified theories* (GUTs).

Unification of the weak and electromagnetic interactions is not manifest until energies of the order of the W and Z masses, approximately 90 GeV/c^2. To get some idea of the energy scale involved in a grand unified theory, we can use the fact that the weak and strong couplings in the standard model are running coupling constants, that is, they vary with energy. When their predicted energy dependences are used to extrapolate the couplings to higher energies, the three curves fail to meet at a point. Nevertheless, this suggests that the interactions are approximately equal at an enormous *unification mass* (M_U) of about 10^{15} GeV/c^2. At the equivalent energy the theory has a single GUT coupling (α_U), with a value of approximately $\frac{1}{42}$.

There are many competing grand unified theories but the simplest incorporates the known quarks and leptons into common families. In a way this is natural, since the fact that the proton and electron have electric charges of the same magnitude hints at a deep relationship between quarks and leptons. In this version of a GUT, one family consists of the three coloured d quarks and a pair of leptons: $(d_r, d_b, d_g, e^+, \bar{\nu}_e)$. Transitions between quarks and leptons are made possible by introducing two new gauge particles, denoted X and Y, with electric charges $-\frac{4}{3}e$ and $-\frac{1}{3}e$, where e is the magnitude of the electron charge. Their masses would necessarily be roughly equal to the unification mass.

This simple model has a number of attractive features. For example, the sum of the electric charges of all the particles in a given family is zero. So, using the multiplet $(d_r, d_b, d_g, e^+, \bar{\nu}_e)$, it follows that $3q_d + e = 0$, where q_d is the charge of the down quark. Thus $q_d = -\frac{1}{3}e$ and the fractional charges of the quarks can be seen to originate from their three colour states. By a similar argument, the charge of the up quark can be found to be $q_u = \frac{2}{3}e$. With the usual quark assignment ($p = uud$), the proton charge can be obtained from $q_p = 2q_u + q_d = e$. So, this GUT gives an explanation for the long-standing puzzle of why the proton and positron have precisely the same electric charge.

All grand unified theories, including this one, make a number of predictions that can currently be tested. If the three couplings of the weak and strong interactions really did meet at a point when extrapolated to high energies, then the three low-energy couplings of the standard model would be expressible in terms of the GUT coupling and the unification mass. This approach can be used to calculate either α_U or M_U, or equivalently the weak mixing angle θ_W. The results are close to the measured value of θ_W but not exactly equal to it. To date, only θ_W has actually been measured.

The exchange of the new gauge bosons would allow transitions between leptons and quarks and so neither baryon nor lepton numbers would be conserved. That would make the proton unstable and open the way for several possible final states, including a pion and a positron ($p \rightarrow \pi^0 + e^+$) and a pion and an anti-neutrino ($p \rightarrow \pi^+ + \bar{\nu}_e$). Examples of Feynman diagrams for the first of these decay modes are shown in Figure 10.1. Interestingly, we already know something about the lifetime of the proton from the existence of life itself. If the proton decayed, the resulting radiation from the material of our bodies could kill us depending, of course, on the rate of radiation produced. For that not to happen, the proton lifetime must be greater than about 10^{16} years, taking account of reasonable uncertainties in the values of g_U and M_U. Grand unified theories give much better bounds than this rough estimate. For the decay modes noted, the predicted value of the proton's lifetime is 10^{30} to 10^{31} years. For comparison, the age of the universe is believed to be about 10^{10} years. Experimenters have looked for proton decay via the two modes specifically mentioned above, for example in the Kamiokande experiment, but no events have yet been observed. This has left physicists to conclude that the proton lifetime is greater than about 10^{32} years, which rules out the simplest GUTs. Still, there are other, more complicated GUTs, that may yet be proven, some of which incorporate the idea of *supersymmetry*.

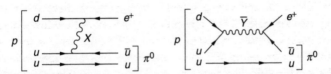

Figure 10.1 Feynman diagrams for the proton decay mode $p \rightarrow \pi^0 + e^+$

Supersymmetry

If there are new particles associated with the unification energy scale, as proposed in GUTs, they would have to be included as additional contributions in higher-order calculations in the electroweak theory – contributions that would upset the delicate cancellations that ensure finite results throughout the standard model. To avoid this upheaval, some means of cancelling these new contributions would need to be found: supersymmetry provides those means.

Supersymmetry (SUSY) is the proposal that every known elementary particle has a *superpartner*, identical to it in all respects except its spin. Spin-$\frac{1}{2}$ particles have spin-0 superpartners and spin-1 particles have spin-$\frac{1}{2}$ superpartners. To distinguish between a spin-$\frac{1}{2}$ particle and its superpartner, an 's' is put before the superpartner's name: for example, a spin-$\frac{1}{2}$ electron has a spin-0 selectron as its superpartner. The full set of elementary particles and their superpartners, in a simple SUSY model, is shown in Table 3. (This is itself a simplification, because even the simplest SUSY model requires three neutral and two charged Higgs bosons.)

If there were an exact correspondence between particles and superparticles, a particle and its superparticle would have identical masses. Such states have not been found so this is clearly not the case. At best, supersymmetry is an approximate symmetry of nature; nevertheless, even in an approximate symmetry the couplings of the two states are equal and opposite, ensuring cancellation, if their masses are not too large. In practice, GUTs that incorporate supersymmetry usually assume that the masses of the superparticles are roughly equal to the masses of the W and Z bosons. With the inclusion of superparticles, there is a slight change in the way the standard model's coupling constants vary with energy; when the constants are extrapolated to very high energies they now meet very close to a single point. As a

Table 3 The particles of the standard model and their superpartners

Particle	Symbol	Spin	Superparticle	Symbol	Spin
Quark	q	$\frac{1}{2}$	Squark	$\tilde{q}$	0
Electron	e	$\frac{1}{2}$	Selectron	$\tilde{e}$	0
Muon	μ	$\frac{1}{2}$	Smuon	$\tilde{\mu}$	0
Tauon	τ	$\frac{1}{2}$	Stauon	$\tilde{\tau}$	0
W–boson	W	1	Wino	$\widetilde{W}$	$\frac{1}{2}$
Z–boson	Z	1	Zino	$\widetilde{Z}$	$\frac{1}{2}$
Photon	γ	1	Photino	$\widetilde{\gamma}$	$\frac{1}{2}$
Gluon	g	1	Gluino	$\tilde{g}$	$\frac{1}{2}$
Higgs boson	H	0	Higgsino	$\widetilde{H}$	$\frac{1}{2}$

result, the unification mass is increased somewhat to about 10^{16} GeV/c^2. At the same time, the value of α_U remains roughly constant and the predicted lifetime of the proton is increased to about 10^{32} to 10^{33} years, just beyond the reach of today's experiments. Likewise, the value of the weak mixing angle is brought into almost exact agreement with the measured value. Whether all this is simply a coincidence is unclear.

Another area where supersymmetry might be tested directly is in the measurement of the *electric dipole moment* (EDM), a quantity that is a measure of the distribution of charges inside a particle and determines how it interacts with an external electric field. EDMs are exactly zero unless parity and time-reversal are both violated. In principle, the violation of *CP* and *T* observed in the decay of some neutral mesons could generate a non-zero EDM, although in the standard model EDMs only appear in higher orders of perturbation theory and are extremely small. In the case of the electron, the calculated value from the standard model is 10^{-11} times smaller than the largest estimate found from experiment.

Supersymmetric theories contain a host of new particles and couplings and much larger symmetry-violating effects can be generated by these theories. Although no unique predictions have been made, a suitable choice of the parameters of supersymmetric theories leads to values for EDMs of a size that could be measured in the not-too-distant future.

To verify supersymmetry, it will be essential to detect superparticles, which will not be easy. To date, experimenters have concentrated on the direct detection of superparticles in reactions. In the simplest SUSY theory, superparticles are produced in pairs, much like strange particles in strong interactions; the decay of a superparticle must produce at least one superparticle in the final state and the lightest such particle will necessarily be stable. Most versions of SUSY theories assume that the lightest particle will be a *neutralino*, $\tilde{\chi}_0$, which is the name given to a mixture of the photino, the neutral Higgsino and the zino.

If this is the case, a possible reaction that could be studied is $e^+ + e^- \rightarrow \tilde{e}^+ + \tilde{e}^-$, followed by the decays $\tilde{e}^\pm \rightarrow e^\pm + \tilde{\chi}^0$, giving the overall reaction $e^+ + e^- \rightarrow e^+ + e^- + \tilde{\chi}_0 + \tilde{\chi}_0$. Like neutrinos, the neutralinos only have weak interactions, so they will be practically undetectable. Any reactions would be characterised by electron-positron pairs $e^+ e^-$ in the final state that carry only a fraction (typically 50%) of the initial energy. Also, because this is not a two-body reaction, these pairs would not be emitted back to back. Although this, and many other reactions, have already been studied, mainly in experiments at LEP, no evidence for the existence of superparticles has yet been found. The negative results have enabled physicists to set lower limits on the masses of neutralinos and sleptons, in the range of 40 to 100 GeV/c^2. Much larger lower limits for the masses of gluinos and squarks have been obtained in experiments using the CDF detector at the Tevatron accelerator. The search for supersymmetric particles will be a major focus of experiments in the LHC accelerator.

String theories

Undeterred by the lack of immediate success with supersymmetry, many bold physicists are attempting to incorporate gravity into ever-grander unified schemes. The problems are mathematically formidable, not least because the infinities encountered are far more severe than those in either QCD or the electroweak theory. Furthermore, there is presently no successful stand-alone quantum theory of gravity analogous to the other two.

An unusual predicament in assessing such theories is the extreme difficulty they have in making unambiguous predictions that can some day be tested by experiment. The theories that have been proposed to include gravity invariably replace point-like elementary particles with tiny one-dimensional quantised *strings* and are formulated in many more dimensions – usually ten, including one dimension of time – than are observed in nature. Unlike the standard model, with its nineteen free parameters, which include the masses of quarks, coupling constants, mixing angles and so on, these theories have a single free parameter: the string tension. Because we live in a four-dimensional world (including time) the extra dimensions have to be *compactified*; that is, reduced to an unobservably small size. It was hoped that compactification would result in the emergence of the standard model as a unique theory, with its nineteen parameters explained in terms of string theory's string tension. This would be a major triumph. Unfortunately, the early optimism about this mathematical scheme has not been sustained. Despite these setbacks, string theory has provided some powerful theoretical tools that have contributed to a better understanding of gauge theories and their relation to gravity.

As we have seen, the standard model uses a conventional particle picture of interactions. Since the structures of the corresponding quantum field theories (such as QCD) are known, it is possible to make unique physical predictions using perturbation theory and the appropriate Feynman rules. In the string formulation, there are five sets of possible Feynman rules, each operating in a ten-dimensional space-time continuum. String theorists have discovered that this leads to a vast landscape of low-energy theories – possibly as many as 10^{500}! – that could result after compactification. Each of these theories would describe a distinct universe, with its own set of fundamental particles, interactions and parameters. Unless there is a method of choosing among the huge range of possibilities, string theories have little or no real predictive power. For this reason, a lively philosophical debate has arisen about whether they should even be considered as scientific theories, since acceptable theories must make unique, experimentally-testable predictions. However, their proponents counter that string theories are being judged by standards that, historically, have not been applied to other emerging theories.

One controversial approach has been to invoke the *anthropic principle*. Roughly speaking, this states that what we can expect to observe is restricted by the conditions necessary for our presence as observers. In other words, the universe is observed to be the way it is because that is the only way it could exist that would allow humans to be in it considering such questions. This somewhat circular-sounding principle has been used to explain the apparent improbable values of some cosmological constants, but is by no means generally accepted as a way forward for string theories. Other theorists believe that a physics-based method for choosing a unique theory will eventually be found. For instance, the internal consistency of string theories in ten dimensions has been shown to imply the existence of higher-dimensional mathematical objects, *branes* (short for membranes). It has been suggested that using branes will allow theorists to construct an even more fundamental theory, involving eleven dimensions, in which all five supersymmetric string theories are unified. This theory has a name – M-theory – although no one knows if the conjecture is true, let alone how to construct such a theory.

Leaving aside their formidable theoretical complexity, string theories are strictly applicable only at energies where gravitational effects become comparable to those of the gauge interactions. Since gravity is so weak, this does not happen until enormous energies are reached, at approximately $10^{19} GeV/c^2$, the *Planck energy*. By wave-particle duality, this means that strings have a size of around 10^{-35} m. The Plank energy is so large that it is difficult to imagine how string theories could be tested by experiments, although some physicists believe that information produced even at the (comparatively) low energies of the LHC, for example the discovery of superparticles, may help to prove them.

At present, the main appeal of string theories lies in their mathematical beauty and naturalness. Although mathematical beauty is a notoriously hard-to-define concept, string theorists have history on their side. A similar consideration of beauty inspired Dirac to persevere with his idea of anti-particles, despite the apparent absurdity of states with total negative energy, and the lack of any supporting evidence. The belief that correct theories are mathematically elegant was also cited by the pioneers of the quark model, pursued in the absence of free quarks, and by those developing the electroweak unification, despite the initial prediction of zero masses for the participating particles. Both were eventually accepted. On the other hand, many short-lived theories were doubtless also considered beautiful by their originators. Needless to say, experimenters will remain sceptical until definite experimental tests can be designed and made: a great challenge for the next generation of particle physicists.

The nature of the neutrino

The neutrino has always been a mystery particle, whose properties have been difficult to study experimentally because it only interacts via the weak interaction. It took about twenty-five years after Pauli postulated it for its existence to be confirmed and a further fifty years to show that its mass was non-zero. Another piece of the neutrino's history is the assumption that it is a *Dirac particle*, described by the Dirac equation. It might seem this has to be true because this equation formed the basis of Dirac's original description of spin-$\frac{1}{2}$ fermions. However, for neutral states, a particle and its anti-particle do not necessarily have to be distinct and this could be the case for neutrinos, as it is for photons. A neutrino that is identical to its anti-particle is called a *Majorana neutrino* and obeys a variation of the original Dirac equation.

The Majorana neutrino may help physicists understand why quarks and leptons have the masses they do. The three known neutrino masses so far observed in experiments have much smaller masses than the other fundamental fermions. A possible explanation has been suggested in the context of extensions of the standard model, in which it is possible for both types of neutrino – the standard Dirac type and the Majorana type – to co-exist. The observed neutrinos, with their very small masses, emerge naturally as a mixture of the two types. However, there is always a price to pay: in this case, it is the prediction that the 'other' combination corresponds to a very heavy neutrino, yet to be discovered.

How might the puzzle of the nature of the neutrino be resolved? The most direct method is to look for reactions that can only occur if the neutrino is its own anti-particle. The most promising of these reactions is neutrinoless double beta decay. In Chapter 1, beta decay was introduced as the most common way that radioactive nuclei decay. This is a weak interaction in which a charged lepton (either an electron or a positron) is emitted together with a single neutrino (or anti-neutrino, as appropriate). In a very small number of nuclei, a second possibility is allowed, whereby two nucleons decay simultaneously, so that two charged leptons and their associated neutrinos are emitted. Because two neutrinos are emitted, double beta decay is a second-order reaction in the weak interaction and so is very rare, with predicted lifetimes for the particles of around 10^{18} to 10^{24} years, depending on the nucleus concerned. Double beta decay can only be observed if the single beta decay process is forbidden by energy conservation and the parent nucleus is stable against decay by emission of alpha particles or photons. In other words, double beta decay is only observable when it is the only allowed decay mode. The decay mechanism is illustrated in the Feynman diagram of Figure 10.2a. Such decays were first directly observed in 1987 and have since been established for ten nuclei.

Neutrinoless double beta decay involves a parent nucleus that decays by the emission of two electrons but no neutrinos. This is forbidden for a Dirac neutrino, because it violates electron lepton number but is allowed if the neutrino is its own anti-particle. The mechanism is shown in Figure 10.2b. Observation of neutrinoless double beta decay would be resounding evidence that the neutrino is its own anti-particle.

Neutrinoless double beta decay can in principle be distinguished from ordinary double beta decay by measuring the energies of the emitted electrons. In single beta decay, energy is carried away by the undetected neutrino, resulting in a continuous spectrum for the energy of the electron – which

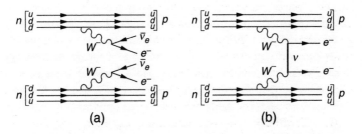

Figure 10.2 (a) Double β-decay, as allowed in the standard model (b) neutrinoless double β-decay, forbidden in the standard model

is why Pauli originally postulated that the neutrino existed. The same is true for the spectrum of the total energy of the two electrons in double beta decay. However, in the neutrinoless case, the electrons carry off all the available energy, which would be observable as a sharp peak in the electron's combined energy. The major problem with mounting an experiment is that the rate for neutrinoless double beta decay is expected to be much smaller even than that for normal double beta decay, perhaps one decay per year (or even fewer) per kilogram of unstable isotope, an incredibly low rate. Because of this, experiments to search for neutrinoless double decays are invariably located deep underground, to protect them from cosmic rays. They must also be shielded to eliminate spurious signals arising from ambient radiation. The sample of the decaying isotope must be extremely pure, because even a very small contamination of a beta-decaying impurity would swamp any signal from double beta decay. Since several kilograms of an isotope are needed to obtain a detectable counting rate, these are very exacting requirements.

One of the experiments currently underway is the NEMO3 detector, located in the Fréjus tunnel beneath Mont Blanc in the French Alps, shown in Figure 10.3. There, a decaying sample of 10 kg of double beta isotopes, in the form of thin sheets, is located in a central tower. The isotopes are surrounded by a detector that can observe and identify the electron tracks from the decays, as well as measure their energies. The detector consists of multi-wire drift chambers to record the electron tracks, electromagnetic calorimeters (plastic scintillator blocks coupled to low-radioactivity photomultiplier tubes) to measure their energy and a magnetic coil to provide a field for charge information. These in turn are surrounded by pure iron shielding, to eliminate gamma rays and wood and water shielding to eliminate neutrons. Overall, the structure is not unlike a conventional detector at a collider (but without the beams!).

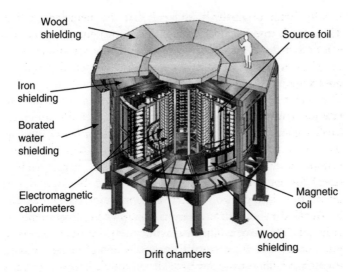

Figure 10.3 Schematic diagram of the NEMO3 detector (courtesy of the NEMO3 collaboration)

NEMO3 and other experiments are still in their early days and they have not yet provided evidence for neutrinoless double beta decay. The most recent results point to an upper limit of 0.5 eV/c² on the effective mass of a neutrino that is its own anti-particle. This is not very different to limits that exist for the masses of Dirac neutrinos.

11

Astro-particle physics and cosmology

Particle physics and astrophysics overlap in an increasing number of areas, with the field of astro-particle physics experiencing a period of rapid growth. The connections between particle physics and astrophysics are particularly important in cosmology, where, for example, neutrinos can provide unique information about topics as diverse as the dynamics of the exploding supernovas, the origin of cosmic rays and even the creation of the universe in the Big Bang. The extreme conditions of temperature and densities present in the early universe implied by the Big Bang model can only be approached, however remotely, in high-energy particle collisions, conducted at energies that are also relevant to the grand unified and SUSY theories of particle physics.

Neutrino astrophysics

Cosmic rays and emissions from the Sun have been key to understanding neutrinos and have led particle physicists to revise their view that neutrinos are strictly massless. In a similar way, ultra high-energy neutrinos could be a rich source of information about galactic and extra-galactic objects, as well as more general cosmology.

One of the first neutrino astrophysics experiments entailed the observation of neutrinos from the explosion of a star, the rare supernova. The chain of events that lead to a supernova starts with a star generating energy from the fusion of light elements into heavier ones. If a star has a mass about eleven times greater than the mass of the Sun, it can evolve through all stages of fusion, ending as a core of iron surrounded by shells of lighter elements. Since no energy is released by the fusion of iron, the core will start to contract under its own weight, by the action of gravity. However, electrons in the dense gas of the core are constrained by the exclusion principle, which means they cannot occupy identical quantum states. Attempts to compress them will result in some electrons being forced to occupy higher energy levels, with larger particle momenta. The resulting force that resists the compression is the *electron degeneracy pressure*.

Initially, this pressure successfully opposes the gravitational contraction, but as more of the outer layers are 'burned' and more iron is deposited in the core, the rise in temperature (or equivalent energy) causes the electrons to become increasingly relativistic. When the core mass reaches about 1.4 times the mass of the Sun, the electrons become ultra-relativistic and the electron degeneracy pressure can no longer balance the inward force of gravity: the star is on the brink of a catastrophic collapse. As the core caves in, the energy of the electrons increases to a point where the weak interaction, $e^- + p \rightarrow n + \nu_e$, becomes possible. Eventually the hadronic matter of the star is predominantly converted into neutrons, a neutron star. The collapse ceases when the gravitational pressure is balanced by neutron degeneracy pressure (an effect of the exclusion principle but applied to neutrons). By then, the radius of the star is typically just a few kilometres. In the case of very heavy stars, neutron degeneracy pressure may be unable to balance the gravitational force and the collapse continues unchecked, resulting in a black hole.

The end of the star's collapse by neutron degeneracy pressure is very sudden. The core material produces a shock wave that travels outwards through the collapsing outer material, creating the supernova explosion. There is an intense burst of electron neutrinos, with energies of a few MeV, which lasts for a few milliseconds. Almost immediately, the core rapidly becomes so dense that it is opaque even to neutrinos. Next, the core material enters a phase where all its constituents (nucleons, electrons, positrons and neutrinos) are in thermal equilibrium. In particular, all flavours of neutrinos will be present, because of the reactions $\gamma \leftrightharpoons \ell^+\ell^- \leftrightharpoons \upsilon_\ell \bar{\upsilon}_\ell$, where ℓ can be any lepton, an electron, muon or tauon. The two-way arrows express the fact that in equilibrium, the reactions will proceed in both directions at equal rates. This process relies on the existence of the Z^0 particle and the weak neutral interaction. Eventually, neutrinos of all flavours, with average energies of about 15 MeV, will diffuse out of the collapsed core and be emitted in all directions over a period that can be calculated to be between 0.1 to 10 s. Taken together, the neutrinos account for about 99% of the total energy released in a supernova. Even so, the immense total energy output means that even the 1% we can see is enough to produce a spectacular, if short-lived, visual effect.

The first experiments to detect neutrinos from a supernova were a version of the Kamiokande experiment and a second experiment, the IMB collaboration, which also used a water Čerenkov detector. Both had been constructed to search for proton decay as predicted by grand unified theories but by good fortune both detectors were operational in 1987 at the time of a spectacular supernova explosion (SN1987A). The Kamiokande team found twelve electron anti-neutrino events and the IMB team eight events, all arriving over the course of approximately ten seconds. Their energies fell in the range 0 to 40 MeV, consistent with estimates of the energies of anti-neutrinos produced in the equilibrium reaction and then diffused from the supernova after the initial pulse. Using the fact that the time of arrival on Earth of a neutrino depends on its initial energy, and hence its mass, physicists deduced that the mass of the anti-neutrino of the electron type is less than 20 eV/c^2. Although this is much larger than the best limit of 2 eV/c^2, obtained from the beta decay of tritium, it is still a remarkable measurement and showed that neutrino astronomy can be used to discover astrophysical information.

The neutrinos from the supernova SN1987A were low energy but there is also great interest in detecting ultra high-energy neutrinos. We know there are gamma ray sources with energies in the TeV range, many of which have their origin within *active galactic nuclei*, compact regions at the centre of galaxies with higher than normal luminosities over some or all of the electromagnetic spectrum. It is an open question whether this implies the existence of sources of neutrinos with similar energies. To find the answer, physicists would need to detect neutrinos travelling upwards through the Earth, as the signal from downward-travelling neutrinos would be swamped by those produced from pion decay in the atmosphere. Like all weak interactions, the detection rate would be very low, especially so for such a rare high-energy event.

To detect neutrinos in the TeV energy range using the Čerenkov effect in water needs huge volumes, many times larger than that used in the SuperKamiokande detector. One ingenious solution involves using the vast quantities of water available in the oceans or frozen as ice at the South Pole; several experiments have been built to take advantage of these sources. The largest so far is the Antarctic Muon And Neutrino Detector Array (AMANDA), at the geographical South Pole. AMANDA (shown in Figure 11.1) consists of strings of optical modules containing photomultiplier tubes that convert the Čerenkov radiation to electrical signals, placed in the ice at great depths using a novel hot water boring device; the ice refreezes around the detectors

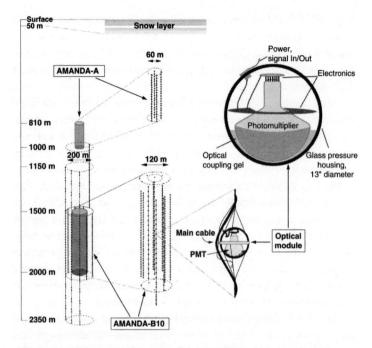

Figure 11.1 Schematic diagram of the AMANDA neutrino detector

in the bored holes. The current version of the detector covers a cylinder with a diameter of 120 m. It has successfully detected atmospheric neutrinos and produced the most detailed map of the high-energy neutrino sky to date but no candidate for a point source has yet been observed.

AMANDA can detect neutrinos with energies up to about 10^{15} eV but an even bigger detector, *IceCube*, is under construction at the South Pole. The detector is expected to be operational in 2011; it will cover an area of 1 km^2 at depths of between 1.4 and 2.4 km, about seventy times the area covered by AMANDA. IceCube will be capable of detecting neutrinos with energies as high as 10^{18} eV. Detection of even higher energy neutrinos is being pursued in several innovative experiments, for example the Antarctic Impulsive Transient Antenna (ANITA), which aims to shed light on the origin of cosmic rays with energies greater than 10^{19} eV.

ANITA exploits a phenomenon similar to the Čerenkov effect as it looks for neutrinos associated with the interaction of cosmic rays with the cosmic microwave photons that pervade the universe and are part of the evidence for the Big Bang. A particle travelling faster than the speed of light in a dense medium transparent to radio-frequency electromagnetic radiation produces a shower of charged particles. These emit a cone of radiation – the *Askaryan effect* – in the radio or microwave part of the electromagnetic spectrum. Because ice is transparent to radio waves up to a frequency of 1.5 GHz, the neutrinos cascading through the Antarctic ice sheet produce a strong electromagnetic pulse that propagates through the ice. Basically, the ice sheet acts as a converter of neutrino energy to radio waves.

The ANITA experiment consists of a detector system mounted on a balloon platform at a height of about 40 km above the Antarctic ice shelf. Driven by the continuous wind circulation around the South Pole, the balloon traverses a circum-polar flight path from which the detector is able to see the ice below it extending out to the horizon at about 700 km distant, effectively a telescope lens that covers an enormous area of approximately 1,500,000 m^2.

The early universe: dark matter and neutrino masses

As we have seen, the modern description of the universe, the Big Bang model, is based on the observation that the universe is expanding, having originated in a very small region of extreme temperatures and densities. Because the distribution of matter in the universe is uniform in all directions at large distances, the expansion appears the same to all observers, irrespective of their locations. Two pieces of evidence for this model are the existence of a cosmic microwave background radiation and the abundance of light elements in the universe. Whether the expansion will continue indefinitely depends on the average density of the universe ρ, because this quantity determines the gravitational forces at play. At present, the critical density ρ_c – below which the expansion will continue indefinitely and above which it will eventually halt and the universe will start to contract – is about five or six nucleons per cubic metre.

Physicists relate the present density of the universe to its critical value by the relationship $\Omega = \rho/\rho_c$, a relative density that helps to describe the prospects for the universe's future. In the most accepted version of the model – the *inflationary big bang model* – there was an early, short, period of very rapid expansion, in which $\Omega = 1$, setting the relative density equal to the critical density. The relative density can be conveniently split into the sum of three components: $\Omega = \Omega_r + \Omega_m + \Omega_\Lambda$, where Ω_r is the contribution due to radiation, Ω_m that due to matter and Ω_Λ a contribution related to the *cosmological constant* (also referred to as *dark energy*), originally included in general relativity by Einstein to ensure the theory was consistent with a static universe. This term was later set to zero when it was found that the universe is expanding. Interest in it has been rekindled by the more recent discovery that the rate of expansion is actually increasing.

Of these terms, only Ω_r has been accurately measured, from an analysis of the microwave background radiation; it is numerically negligible. The total matter contribution can be deduced from the gravitational energy needed for consistency with observations on the motions of large-scale structures, such as galaxies and clusters of galaxies, in the universe. The value obtained for Ω_m is in the range 0.24 to 0.30. The dark energy term,

Ω_Λ, can be estimated from various cosmological observations, including recently-discovered minute temperature fluctuations in the microwave background radiation; its value is about 0.7. Adding these contributions, we see that the value of Ω is around 1, although the uncertainties involved are considerable. Analysis of data on fluctuations in the microwave background from a recent satellite experiment supports this conclusion and gives a value of 1 with an uncertainty of only 2%. An unsatisfactory feature of this analysis is that the origin of the largest term, the dark energy, is totally unknown.

The matter term, Ω_m, has contributions from a number of sources. That part due to baryons, Ω_b, may be inferred from a knowledge of how nuclei are formed in the universe. Its value is about 0.05. It follows that most matter is non-luminous and baryons are the source of only a small fraction, about 15–20% of the total matter contribution. There could be other sources of non-luminous baryonic matter, for example in the form of brown dwarf stars, which have a mass below that necessary to maintain nuclear fusion reactions, and relatively small black holes, about the size of planets. There is experimental evidence that such *massive compact halo objects* (MACHOs) exist, but in unknown quantities. However, it is not thought that they alone can account for the missing matter of the universe. We are forced to the rather startling conclusion that the bulk of matter, as much as 85%, is non-baryonic, collectively called *dark matter*.

There are several candidates for dark matter. Massive neutrinos might be one possibility. Such particles would have to be heavy enough to have been non-relativistic in the early stages of the universe (*cold dark matter*), because if they were relativistic (*hot dark matter*) they would have rapidly dispersed, giving rise to a uniform energy distribution in space. Calculations suggest that in this case, there would have been insufficient time for the observed galaxies to have formed. The contribution of massive neutrinos can be calculated once the number of neutrino types is known and from the size of the dark matter term it is clear that the sum of the masses of all possible neutrino types cannot exceed $10–12$ eV/c^2.

These values are large and not very useful but fortunately much better limits can be obtained from other sources. This is because neutrinos with masses as small as 0.1 eV/c^2 can have an observable effect on the formation of large-scale structures in the universe by dampening the growth of the perturbations that eventually lead to such structures. Two major satellite surveys of these sorts of large-scale structures have been conducted, each of which has provided limits for the sum of the masses of all neutrinos. Some astrophysicists claim a value as low as 0.2 eV/c^2 but a more conservative estimate is $0.5–1.0$ eV/c^2. Even allowing for some statistical uncertainty, this is still lower than the bound from tritium decay. It should be noted that this range cannot strictly be viewed as a direct measurement, since it depends on the details of the cosmological model used. However, if these analyses are correct, they imply that neutrinos play a minor role in the matter deficit of the universe.

It is now believed that the most important constituent of cold dark matter comes in the form of *weakly interacting massive particles* (WIMPs). Scientists have not yet found any particles that have the required properties, but the most likely candidates are particles assumed to exist in supersymmetric models, particularly the lightest such state, usually taken to be the neutralino, although it is not the only possibility. Experiments such as AMANDA can search for WIMPs, even though they were not primarily designed for this.

Recently, several dedicated experiments have been mounted to detect the recoil energy, about 50 keV, of nuclei when they are struck by WIMPs. In principle, such recoils can be identified in a number of ways. For example, in semi-conducting materials such as gallium arsenide, free charge will be produced that can be detected electronically; in a scintillator, such as sodium iodide, the emission of photons can be collected using photomultipliers;

and in crystals at low temperatures, energy can be converted to microscopic vibrations that can be detected as a very small rise in temperature.

In practice, scientists have struggled to overcome the very low expected detection rate that is predicted, based on the expected velocities and assumed masses of WIMPs. If WIMPs are taken to be neutralinos, expectations range from one to ten events per kilogram of detector per week, a much smaller rate than that from naturally-occurring radioactivity, including that from the material of the detectors themselves. Like the experiments searching for neutrinoless double beta decays, those searching for WIMPs are located deep underground, to shield the detector from cosmic rays, and in areas without radioactive rocks. The detectors themselves are built using extremely pure materials, to minimise radioactive signals. These experiments are at an early stage but have already ruled out some versions of supersymmetric theories with low-mass neutralinos.

Quark-gluon plasma

In ordinary matter, quarks are confined within hadrons. However, at extremely high energy densities, conditions would begin to approximate the earlier stages in the universe during which quarks and gluons were free to move across a volume larger than that of a hadron. Approximate calculations in QCD suggest that this should occur at an energy density about six times that at the centre of a heavy nucleus, corresponding to an effective temperature of 10^{12}–10^{13} K. This would be a totally new state of matter; a *quark-gluon plasma*. This name draws upon the fact that the interiors of stars contain a plasma of electrons and ions. A quark–gluon plasma would have existed in the first few hundredths of a second after the Big Bang and may exist now in the centre of neutron stars. Several searches for this new state of

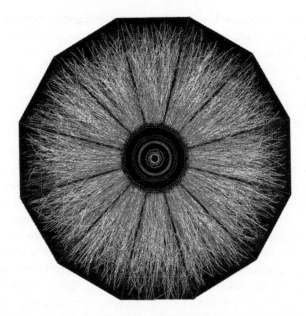

Figure 11.2 Particles produced in a 200 GeV gold-gold interaction in a time projection chamber at the RHIC

matter have been spearheaded by research teams at the Relativistic Heavy Ion Collider (RHIC) at Brookhaven National Laboratory. This machine collides two counter-circulating beams of gold ions with energies of 200 GeV per nucleon, which produces several thousand particles in the final state. An example recorded in a time projection chamber at the RHIC is shown in Figure 11.2.

The RHIC is being used to investigate whether the energy density in such collisions is sufficient to produce a quark-gluon plasma and, if so, how the plasma behaves as it cools. These questions can be answered by studying the rates of production and the angular distributions of the various particles produced in the collider. The most recent results indicate that a quark–gluon plasma is indeed formed. Studying the properties of the plasma should help scientists understand how quarks become deconfined.

The asymmetry of matter and anti-matter

One of the most striking facts about the universe is the paucity of anti-matter, compared to matter. Cosmic rays are overwhelmingly composed of matter; what little anti-matter is present is easily explained as the product of intergalactic collisions between matter and photons. Neither do astrophysicists see intense outbursts of electromagnetic radiation that would accompany the mutual annihilation of regions of matter with regions of anti-matter. The absence of anti-matter is completely unexpected, because in the original Big Bang it would be natural to assume a total baryon number of zero. Although in principle this is not necessary, most physicists would be very uneasy supposing that such a lack of basic symmetry existed at the start of the universe.

Continuing with this assumption, during the period when the effective temperature was large compared to hadron energies, baryons and anti-baryons would be in equilibrium with photons via reversible reactions such as $p + \bar{p} \leftrightarrows \gamma + \gamma$. These reactions would continue until the temperature fell to a point where the photons no longer had sufficient energy to produce $p\bar{p}$ pairs and the expansion had proceeded to a point where the density of protons and anti-protons was such that their mutual annihilation became increasingly unlikely. The ratios of baryons and anti-baryons to photons would 'freeze' to a value around 10^{-18}, with equal numbers of baryons and anti-baryons. While these ratios would in theory continue to be maintained over time, the actual observed ratios are approximately 10^{-9} for baryons to photons and 10^{-13} for anti-baryons to photons, implying that the ratio of anti-baryons to baryons is about 10^{-4}. Here the simple Big Bang model fails spectacularly.

The conditions whereby a baryon–antibaryon asymmetry could arise were first stated by the Russian physicist Andrei Sakharov. The asymmetry requires first, an interaction that violates baryon number; second, an interaction that violates charge conjugation C and the combined symmetry CP; and third, a non-equilibrium situation at some point that can seed the process. Earlier, I discussed the evidence that CP is violated in the decays of some neutral mesons but the source and size of this violation are not compatible with what is required to account for the observed baryon-antibaryon asymmetry. The conclusion must be that there is another, as yet unknown, source of CP violation. One possibility is the additional richness of CP-violating effects present in supersymmetric theories. Likewise, a method for generating a non-equilibrium situation is also unknown, although it may be that the interactions that violate baryon number in grand unified theories, necessary for the first condition, could provide the answer. The asymmetry of matter–antimatter in the universe remains a serious unsolved problem to be tackled.

The matter-antimatter asymmetry is only one of a number of problems that will be intensively studied in the coming years. It will be particularly important to test the assumptions of supersymmetric models and to determine whether the host of new particles and couplings they predict do exist. If confirmed, this could not only explain the matter-antimatter asymmetry but could also lead to a better understanding of cosmology in general, including the central question of how the universe began.

Afterword

We cannot but admire the impressive progress that the young field of particle physics has made in a mere century. In particular, the last forty or so years have witnessed the rise of the standard model from the interplay of theoretical speculation and ingenious experiments and the great strides in the use of that model to interpret the strong and electroweak interactions of particles. Taken together with the general theory of relativity, the standard model is capable of describing phenomena ranging from the interactions of the smallest known objects to the dynamics of the largest structures in the universe, a true tribute to the scientific method.

Table 4 provides a succinct summary of the particles of the standard model and their interactions which we have discussed. Establishing the existence of the elusive Higgs boson, the only particle in the table not yet confirmed, will complete our understanding of the electroweak theory and the origin of mass and provide a firm foundation for extending the standard model. Should the Higgs particle not be found in the experiments at the LHC at CERN, or the Tevatron at Fermilab, we will need to have a major re-think about these questions.

The successes of the standard model have spurred physicists to look deeper, investigating why there are only three generations of quarks and leptons, why the forces of nature have different strengths and why gauge invariance is a singular property for acceptable quantum field theories. Attempts to answer these questions have led physicists to build ambitious constructs that

Table 4 Particles of the standard model and their interactions. *The notation for the forces is: s (strong), em (electromagnetic), w (weak), g (gravity). Antiparticles are not shown*

Type	Symbol	Name	Force			Charge	
			Experiences	Mediates	Spin	Electric	Colour
leptons	e, μ, τ	electron, muon, tauon	g, w, em		$\frac{1}{2}$	-1	no
	ν_e, ν_μ, ν_τ	electron neutrino etc	g, w		$\frac{1}{2}$	0	no
quarks	u, c, t	up, charm, top	g, w, em, s		$\frac{1}{2}$	$+\frac{2}{3}$	yes
	d, s, b	down, strange, bottom	g, w, em, s		$\frac{1}{2}$	$-\frac{1}{3}$	yes
bosons	γ	photon	g, w, em	em	1	0	no
	W	charged gauge bosons	g, w	w	1	± 1	no
	Z	neutral gauge boson	g, w	w	1	0	no
	g	gluons	g, s	s	1	0	yes
	H	Higgs boson	g, w		0	0	no

encompass more forces in a single unifying theory. As the experiments at the LHC produce results, the predictions from these grand unified and supersymmetric theories will be tested and the outcomes will help to solve outstanding problems, such as why there is so little anti-matter in the universe.

The most ambitious theories coming out of particle physics today strive to unite the strong and electroweak interactions with gravity using a new conception of particles, imagining them as vibrations of one-dimensional strings. The prospects for making unique predictions from string theories are at present limited, as are the chances of testing any of these predictions experimentally. It will be a great challenge for string theorists to derive unique predictions that can be tested at present energies. Because current tests often require energies approaching those that existed at the time of the Big Bang, particle physicists, astrophysicists and cosmologists are more and more united in solving many common problems, including the search for possible superparticles and for an explanation of dark matter and the mysterious origin of dark energy. However, not all the outstanding questions about the universe and its particles need huge accelerators; supersymmetric theories make predictions about electric dipole moments, the stability of the proton and the nature of the neutrino that can be tested in smaller experiments and will, physicists hope, lead to more breakthroughs.

The short history of particle physics has been continual discovery and understanding. What does its future hold? Specific predictions cannot be guaranteed, but we can definitely say that there are immensely exciting prospects for future research and undoubtedly surprises in store, as scientists continue to seek answers to questions about the nature and very origins of the universe. The advice given to the young Max Planck not to study science because there was 'nothing left to discover' is as erroneous today as when he heard it in 1875.

Glossary

Accelerator Machine that uses electric fields to accelerate charged particles to high energies.

Alpha particle (α) Nucleus of the helium atom; consists of two protons and two neutrons bound together by the strong nuclear interaction; emitted in some decays of radioactive nuclei.

Angular momentum Property of rotational motion; contributes to the energy of the rotating particle. A quantum particle may possess both orbital angular momentum, by virtue of its rotation about a fixed axis and an intrinsic spin angular momentum, the latter present even when the particle is at rest.

Anti-matter Every particle of matter has a corresponding particle of anti-matter characterised by the opposite sign of its internal quantum numbers but the same spin and mass; particles and their anti-particles can mutually annihilate and in the process convert their mass to energy.

Anti-particle Anti-matter version of a particle. Examples: anti-proton, anti-quark. In a few cases, a particle and its anti-particle are identical, as with the photon.

Asymptotic freedom Reduction in the strength of the strong force between quarks as the distance between them decreases.

Atom System of matter consisting of a number of electrons surrounding a very small, positively charged central nucleus; the smallest unit that can be identified uniquely as a chemical element.

***B*-factory** Colliding beam accelerator whose energy is tuned to produce copious quantities of particles containing bottom quarks or anti-quarks.

Baryon Class of strongly interacting particles consisting of three quarks. Examples: nucleons, hyperons.

Baryon number Internal quantum number characterising baryons. The total baryon quantum number of a system of particles is the sum of the number of baryons minus the number of anti-baryons.

Beta (β) decay Transition caused by the weak interaction, characterised by the emission of an electron (or positron) and a neutrino.

Big Bang Event that marked the start of the universe, when it rapidly expanded from a small region of enormous density and temperature.

Black–body radiation Electromagnetic radiation emitted by all heated bodies; has a distinct distribution of wavelengths determined by the temperature of the body.

Boson Particle with an integer value of spin, such as photon, gluon or pion.

Bottom Internal quantum number characterising the bottom quark. The total bottom quantum number of a system of particles is the sum of the number of bottom quarks minus the number of anti-bottom quarks.

Bottomium Composite system of a bottom quark and an anti-bottom quark.

Bottom quark (*b* quark) Heavy quark with electric charge $-\frac{1}{3}$, in units of the magnitude of the charge on the electron.

Bubble chamber Particle detector that shows particle trajectories via a trail of bubbles in a superheated liquid; in operation in the 1950s and 1960s but now obsolete.

Cabibbo angle Parameter that determines the mixing between strange and down quarks and their role in the weak interaction.

CERN European Centre for Nuclear Research (actually particle physics research) in Geneva, Switzerland.

Charge conjugation Operation whereby all particles are changed to their anti-particles; leads to a quantum number C that is conserved in the strong and electromagnetic interactions but not in the weak interaction.

Charged weak interactions Weak interaction mediated by a W gauge boson.

Charm Internal quantum number characterising the charm quark. The total charm quantum number of a system of particles is the sum of the number of charm quarks minus the number of anti-charm quarks.

Charmonium Composite system of a charm quark and an anti-charm quark.

Charm quark (*c* quark) Heavy quark with electric charge $+\frac{2}{3}$, in units of the magnitude of the charge on the electron.

Čerenkov radiation Electromagnetic radiation emitted by a charged particle travelling in a medium with a speed greater than the velocity of light in that medium.

Classical physics Physical theories prior to the development of quantum theory.

Collider Particle accelerator where two beams of charged stable particles moving in opposite directions collide head-on.

Colour Internal quantum number of quarks; source of the fundamental strong interaction analogous to electric charge as the source of the electromagnetic interaction.

Conservation law Law stating that a specific quantity, for example energy, is the same before and after an interaction.

Cosmic rays High-energy particles, mainly protons, impinging on Earth's atmosphere from space.

Coupling constant Number measuring the intrinsic strength of the interaction between particles; in general, varies with energy.

CPT theorem Statement that all relativistic quantum field theories are unchanged (invariant) under the combined operations of charge conjugation, parity and time-reversal.

Cyclotron Early form of cyclic accelerator; no longer used in particle physics.

Dark matter Matter that is postulated to explain the fact that there is more matter in the universe than has been directly observed; candidates include heavy neutrinos, neutralinos and WIMPs.

Decay A situation whereby an unstable system transforms to a more stable system with a smaller energy.

Deep inelastic scattering High-energy interaction of a lepton with a target nucleon where the lepton penetrates deep into the structure of the nucleon.

Dirac equation Equation describing the electron that incorporates relativity and quantum theory; led to the prediction of anti-particles and an understanding of spin.

Double beta decay Beta decay with the emission of two electrons and two neutrinos.

Down quark (_d_ quark) Light quark with electric charge $-\frac{1}{3}$, in units of the magnitude of the charge on the electron.

Eightfold Way Classification scheme for hadrons, which led to the quark model.

Electric dipole moment (EDM) Quantity that depends on the distribution of electric charge in a particle and determines how it interacts with an external electric field; non-zero only if time-reversal is not conserved.

Electromagnetic interaction Interaction between electrically charged particles, mediated by the exchange of photons.

Electromagnetic radiation Radiation emitted by all accelerating charged particles.

Electron (e^-) Stable, negatively-charged constituent of atoms; member of the family of leptons.

Electroweak interaction Theory uniting the weak and electromagnetic interactions.

Electron–volt Amount of energy that an electron gains when accelerated through a potential of one volt

$E = mc^2$ Equation expressing the relationship between the energy and mass of a particle in terms of the speed of light in a vacuum.

Fermion Generic term for any particle with half-integer value of spin, such as the proton or electron.

Feynman diagram Pictorial representation of particle interactions; leads to numerical values for measurable quantities in particle physics using perturbation theory.

Field Region of space-time that has been assigned some physical property.

Fine structure constant Coupling constant of electromagnetic interaction.

Fission Break-up of heavy nucleus to smaller nuclei.

Flavour Generic name for the quantities (internal quantum numbers and masses) that distinguish one type of quark from another. For example, up and down quarks have different flavours.

Fusion Combination of light nuclei to form heavier nuclei.

Gamma rays Photons with energies in the GeV region.

Gauge bosons Bosons that transmit the fundamental electroweak and strong forces.

Gauge theories Theories that possesses a property whereby certain mathematical manipulations of the theoretical quantities that appear in the theory may be made but which leave the physical predictions of the theory unchanged, such as quantum electrodynamics, quantum chromodynamics.

Generation Classification of the families of quarks and leptons. There are three generations of each and each generation consists of two quarks and two leptons (one charged and its associated neutrino).

GeV 10^9 electron volts.

Glueballs Hypothesised composite particles consisting only of gluons.

Gluon Spin-1 'force carrier' of the fundamental strong interaction analogous to photons for the electromagnetic interaction; electrically neutral but with non-zero colour quantum number.

Grand unified theories (GUTs) Theories that attempt to unite the electroweak and strong interactions.

Gravity The feeblest of the four forces of nature; experienced by all matter.

Hadron Composite of quarks that experiences the strong interaction.

Higgs boson Massive spin-0, neutral particle that is the source of mass in the standard model of electroweak interactions.

Higgs mechanism Mechanism whereby massless particles acquire masses by interacting with the Higgs field without destroying the gauge invariance of the interaction.

Hyperon Unstable baryon with non-zero strangeness, whose decay products include either a nucleon or another hyperon.

Infra-red slavery Increase in the strength of the strong force between quarks as the distance between them increases. Leads to confinement of quarks in hadrons.

Ion Atom carrying a positive, or negative, electric charge as a result of having had an electron removed, or added, to it.

Jets System of particles produced in reactions at high energies and grouped around a fixed direction; originating from quarks and gluons.

J/ψ particle First member of the charmonium family to be discovered.

Kaon (K meson) First meson to be discovered with a non-zero strangeness quantum number.

keV 10^3 electron volts.

Kinetic energy Energy due to a body's motion.

LEP Large Electron Positron collider at CERN.

Lepton Spin-$\frac{1}{2}$ particle that does not experience the strong interaction, such as the electron, muon and neutrino.

Lepton number Quantum number associated with a lepton generation; conserved in all interactions.

LHC Large Hadron Collider accelerator at CERN.

Linac Linear accelerator.

MACHO MAssive Compact Halo Object such as brown dwarf star, small black hole.

Majorana neutrino Type of neutrino that is its own anti-particle.

Magnetic moment Quantity that determines the behaviour of a particle in the presence of a magnetic field.

Mass Quantity of matter that characterises a body and a measure of its resistance to accelerating forces. The mass of a body is invariant, whereas its weight depends on the gravitational forces present.

Meson Type of hadron; composite of a quark and an anti-quark.

MeV 10^6 electron volts.

Mixing A property in quantum theory that allows a set of states to be written as a linear combination of another set of other states, without altering any physical predictions of the theory.

Molecule Cluster of atoms bound together by the electromagnetic force.

Muon (mu) Heavier unstable sibling of the electron; decays via the weak interaction.

Neutrino Electrically neutral lepton; exists in three distinct varieties (flavours).

Neutrinoless double decay Beta decay with the emission of two electrons but no neutrinos; only allowed for Majorana neutrinos.

Neutrino mixing (see Mixing).

Neutrino oscillations Possibility of neutrinos of one type transforming into neutrinos of another type as a consequence of mixing; only possible if neutrinos have non-zero masses.

Neutral weak interactions Weak interaction mediated by a Z boson.

Neutron Electrically neutral constituent of the nucleus; about 0.1% heavier than the proton.

Nucleon Generic name for protons and neutrons, the constituents of nuclei.

Nucleus Positively-charged small dense core of atoms.

Parity The operation of reflecting all spatial co-ordinates; leads to a quantum number P that is conserved in strong and electromagnetic interactions but not in the weak interaction.

Periodic table Table of chemical elements ordered by the number of protons in the atomic nucleus, to highlight similar properties.

Perturbation theory Method of calculation whereby the dominant contribution to a process is modified by progressively smaller terms.

Photon Massless spin-1 boson; mediator of the electromagnetic interaction.

Pion Lightest meson; exists in three charged varieties.

Planck energy Energy where the gravitational force becomes comparable to those of the gauge interactions; approximately 10^{19} GeV.

Planck's constant (h) Very small fundamental constant that plays a major role in determining the nature of the quantum world. For example, the spins of particles are half integer multiplets of $h/2\pi$.

Positron (e^+) Anti-particle of the electron; has positive electric charge.

Proton (p) Constituent of the nucleus with positive electric charge.

Quantum chromodynamics (QCD) Theory of the strong interaction between coloured quarks mediated by the exchange of coloured gluons.

Quantum electrodynamics (QED) Theory of the electro-magnetic interaction mediated by the exchange of photons.

Quark–gluon plasma State of matter consisting of quarks and gluons freely moving in a volume large compared to that of a single hadron; believed to exist at effective temperatures above approximately 10^{12} K.

Quarks Elementary spin-$\frac{1}{2}$ particle of the standard model; exists in six different flavours: up, down, strange, charm, bottom and top; the first five are constituents of hadrons.

Radioactivity Term describing the spontaneous decays of some nuclei.

Renormalisation Scheme for removing the infinities that appear in quantum field theories.

Resonance Unstable particle that decays by either the strong, electromagnetic or weak interaction.

Running coupling Any particle coupling that varies with energy.

SLAC Stanford Linear Accelerator Centre, Stanford, California, USA.

SNO Sudbury Neutrino Observatory. An underground laboratory for neutrino physics in Sudbury, Ontario, Canada.

Spin Intrinsic angular momentum of particles; given in multiples of half integer units of Planck's constant, h.

Spontaneous symmetry breaking A form of symmetry breaking, where the gauge invariance of the interaction remains intact.

Standard model Current theory of elementary particle physics consisting of the strong interactions of quarks via the exchange of gluons and the electroweak interaction of quarks and leptons by the exchange of photons, W bosons and Z bosons.

Strange particles Particles having non-zero strangeness quantum number; hence containing one or more strange quarks or anti-quarks.

Strange quark (s quark) Quark with electric charge $-\frac{1}{3}$, in units of the magnitude of the charge on the electron; mass lies above that of the down quark and below that of the bottom quark.

Strangeness Property posed by particles containing at least one strange quark.

String theory Theory of particles based on them being vibrations of a one-dimensional string.

Strong interaction One of the four fundamental forces of nature; binds quarks and anti-quarks within hadrons.

Strong nuclear force Force between hadrons that binds nucleons within nuclei; residual effect of strong force between quarks.

SuperKamiokande Underground neutrino facility located in the Japanese Alps.

Superparticles Particles postulated in supersymmetric theories to 'match' existing particles, such as selectrons, gluinos and squarks.

Supersymmetry (SUSY) Theory uniting fermions and bosons where every known partner is matched by another partner, yet to be discovered, differing from it by a spin of $\frac{1}{2}$.

Symmetry Term used to describe the situation where a system or theory is unchanged (invariant) when certain operations are performed on it. Example: a circle has rotational symmetry.

Synchrotron Type of circular accelerator.

Tauon (tau) Heaviest lepton.

Time reversal Operation of reversing all time co-ordinates.

Top quark (t quark) Heaviest quark; has electric charge $+\frac{2}{3}$, in units of the magnitude of the charge on the electron.

Uncertainty principle Statement in quantum theory that energy conservation may be violated, but only for a short time that is given in terms of Planck's constant h.

Unified theories Theories that attempt to combine the strong, electromagnetic and weak interactions and, ultimately, gravity.

Upsilon (Υ) Bound state of a bottom quark and its anti-quark.

Up quark (*u* quark) Light quark with electric charge $+\frac{2}{3}$ in units of the magnitude of the charge on the electron.

Vector meson A meson having unit spin, such as the photon.

Virtual particle Particle that is exchanged between other particles but does not appear in the initial or final state.

W boson Charged spin-1 particle mediating charged weak interactions.

Weak charged interactions Weak interactions due to the exchange of the W boson.

Weak interaction One of the four fundamental forces of nature; mediated by the exchange of W and Z bosons.

Weak neutral interactions Weak interactions due to the exchange of the Z boson.

WIMP Weakly Interacting Massive Particle. Possible candidate is the neutralino, the lowest mass particles in supersymmetric theories

Z boson Neutral spin-1 particle mediating neutral weak interactions.

Further reading

These are some texts that I have found interesting and may be useful for readers wanting to explore particle physics further.

Fritzsch, H. (1983) *Quarks: the Stuff of Matter*. London: Allen Lane
An account of the formulation of the quark model, focused largely on the area of strong interactions. Inevitably somewhat dated now and not always consistent in its assumptions about the prior knowledge of the reader, but still worth reading.

Weinberg, S. (1992) *Dreams of a Final Theory*. New York: Pantheon Books
A classic, elegantly written, personal view of particle physics by Steven Weinberg, one of its leading modern contributors, which includes his vision for its future.

Kane, G. (1995) *The Particle Garden*. Reading, Mass.: Addison-Wesley Publishing Company
A non-mathematical account of particle physics, with some intriguing sections on extensions of the standard model and what is meant by 'understanding' something in science.

Johnson, G. (2000) *Strange Beauty*. London: Jonathan Cape
An excellent biography of Murray Gell-Mann. The story of Gell-Mann's life is woven around clear, non-technical accounts of the many discoveries in particle physics to which he made significant contributions.

Close, F., Marten, M. and Sutton, C. (2002) *The Particle Odyssey*. Oxford: Oxford University Press

A descriptive history of the discovery of all the major particles. Probably too detailed for a beginner but it is lavishly illustrated with many stunning colour photographs.

Close, F. (2007) *The New Cosmic Onion: Quarks and the Nature of the Universe*, 2nd edn. Abingdon, Oxon.: Taylor and Francis

A short introduction to particle physics for a reader with some formal background in science, perhaps a senior school student or junior undergraduate.

Martin, B.R. and Shaw, G. (2008) *Particle Physics*, 3rd edn. Chichester, Sussex.: J. Wiley & Sons

An introductory textbook for a first course in particle physics at undergraduate level.

Dosch, H.G. (2008) *Beyond the Nanoworld: Quarks, Leptons and Gauge Bosons*. Wellesley, Mass.: A.K. Peters Ltd

A succinct review of particle physics, emphasising the origins of modern theories but with only a brief discussion of how experiments are performed. The general reader may find the details quite hard going, because sophisticated mathematical concepts are often used.

Farmelo, G. (2009) *The Strangest Man*. London: Faber and Faber

A superbly written biography of Paul Dirac, containing very clear explanations of those aspects of quantum theory relevant to particle physics to which he contributed.

Index

A Beginner's Guide to Genetics

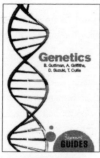

9781851683048
£9.99/ $14.95

With genetics and genetic engineering receiving almost daily coverage in the media, this book is an introduction for general readers who wish to know more about a science that is changing our world. Starting with the history of genetics, from primitive breeding programmes to Mendel's Law, and moving on to a full explanation of its role in our future, this is a comprehensive survey of genetics past, present and future.

"What will make this book so interesting to the general reader is its emphasis on the societal and historical context of each concept discussed. All in all, an excellent resource. Highly recommended!" *Choice Magazine*

BURTON GUTTMAN is a Professor Emeritus of Biology at The Evergreen State College, Washington, USA. Anthony Griffiths is Professor of Botany at the University of British Columbia, Canada.

DAVID SUZUKI is Professor Emeritus at the University of British Columbia, Canada;

TARA CULLIS taught at Harvard University for five years, and is an expert on science and social issues

Browse further titles at
www.oneworld-publications.com

A Beginner's Guide to History of Science

9781851686810
£9.99/ $14.95

Sean Johnston weaves together intellectual history, philosophy, and social studies to offer a unique appraisal of the nature of this evolving discipline. This book demonstrates that science is a continually evolving activity that both influences and is influenced by its cultural context.

"Lucidly and engagingly written ... Johnston has managed to cover an impressive range of material, making it readily accessible to newcomers." **Patricia Fara** – author of *Science: A Four Thousand Year History*

"Clearly written without being patronising, this is a first-rate introduction to the history of science! " **Dr Peter Morris** – Head of Research at the Science Museum, London

SEAN F. JOHNSTON is Reader in the History of Science and Technology at the University of Glasgow. He is also a Fellow of the Higher Education Academy with a prior career as a physicist and systems engineer.

Browse further titles at
www.oneworld-publications.com

A Beginner's Guide to Philosophy of Science

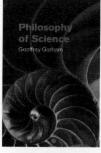

Geoffrey Gorham considers explores the social and ethical implications of science by linking them to issues facing scientists today: human extinction, extraterrestrial intelligence, space colonization, and more.

9781851686841
£9.99/ $14.95

"Lively, accessible, and clear-headed. Good for the beginning student and for anyone wishing guidance on how to start thinking philosophically about science."
Helen Longino – Clarence Irving Lewis Professor of Philosophy at Stanford University

GEOFFREY GORHAM has been teaching and researching philosophy of science for 15 years, and is currently Associate Professor of Philosophy at Macalester College in St. Paul, Minnesota.

A Beginner's Guide to Quantum Physics

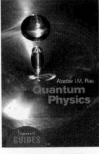

9781851683697
£9.99/ $14.95

From quarks to computing, this fascinating introduction covers every element of the quantum world in clear and accessible language. Drawing on a wealth of expertise to explain just what a fascinating field quantum physics is, Rae points out that it is not simply a maze of technical jargon and philosophical ideas, but a reality which affects our daily lives.

"Rae has done an impressive job. Any reader who is prepared to put in a little effort will come away from this book with an understanding of the basics of some important practical applications of the theory and some appreciation of why its conceptual foundations are still the subject of such spirited debate."
Professor Anthony Leggett – Winner of the 2003 Nobel Prize for Physics

"Rae's emphasis on the practical impact of abstract concepts is very welcome."
Professor Sir Michael Berry – Royal Society Research Fellow, Bristol University

ALASTAIR RAE is editor of *The European Journal of Physics* and was Reader in Quantum Physics at the University of Birmingham until his recent retirement.

Browse further titles at
www.oneworld-publications.com

A Beginner's Guide to Psychology

978-1-85168-578-3
£9.99/ $14.95

From TV experts to the amateur musings of your best friend, the language of psychology has permeated all aspects of every-day life. This Beginner's Guide is informed by the latest cutting-edge research and provides a vibrant and witty examination of the very heart of what it is to be human.

Can personality and intelligence be measured?

Is being physically attractive an advantage?

Is it really better to forgive and forget?

How do babies learn to perceive and think?

Can listening to Mozart improve IQ?

What happens when we sleep?

DR G. NEIL MARTIN is Reader in Psychology, a Fellow of the RSA, a Chartered Scientist, and Director of the Human Olfaction Laboratory at Middlesex University, UK. He has written several books on psychology and neuroscience, and co-authored the first online course in introductory psychology in Europe,

A Beginner's Guide to The Brain

978-1-85168-594-3
£9.99/ $14.95

Using the very latest research in neuroscience this lively introduction explains how 1.4kg of wet grey tissue can not only control all of our bodily functions, thoughts, and behaviours, but also house the very essence of who we are.

"A virtuoso performance! The book is technical, easy to read and entertaining." **Garth Nicholson** – Associate Professor of Medical Genetics, University of Sydney

"A concise primer that summarizes the fascinating complexity of the brain in a memorable and refreshingly graspable manner." **Robert Brown** – Professor of Neurology, Harvard Medical School

AMMAR AL-CHALABI is an Honorary Consultant Neurologist at King's College Hospital and Senior Lecturer at King's College London.

MARTIN R. TURNER is a Specialist Registrar in Neurology at the John Radcliffe Hospital in Oxford, and a Clinical Tutorial Fellow at Green College, Oxford University.

R. SHANE DELAMONT is a Consultant Neurologist at King's College Hospital.

Browse further titles at
www.oneworld-publications.com

Beginners
GUIDES

A Beginner's Guide to Cloning

9781851685226
£9.99/ $14.95

Would you drink milk from a cloned cow? Should we clone extinct or endangered species? Are we justified in using stem cells to develop cures? When will we clone the first human? Ever since Dolly the sheep, such questions have rarely been far from the public consciousness. Aaron Levine explains the science of cloning and guides readers around the thorny political and ethical issues that have developed.

"I highly recommend this book to everyone: I think it would be especially good as additional reading material for introductory genetics courses, bioethics and biomedical classes." *Scienceblogs.com*

AARON LEVINE is currently conducting research on the impact of public policy on biomedical research at the Woodrow Wilson School of Public and International Affairs, Princeton University.

Browse further titles at
www.oneworld-publications.com

A Beginner's Guide to Evolution

9781851683710
£9.99/ $14.95

Burt Guttman assumes no prior scientific knowledge on the part of the reader, and explains each of the ideas and concepts of evolution. Looking ahead to the future of evolutionary theory, and assessing its implications for the way we understand morality, human nature and our place in the world, this book provides the perfect starting point for understanding what evolution is and why it matters.

"Another book in this excellent series that explains the necessary background to understanding evolution."
Scientific and Medical Network

BURT GUTTMAN is Professor of Biology at The Evergreen State College, Washington. His previous works include *Genetics: A Beginner's Guide*, also published by Oneworld.

Browse further titles at
www.oneworld-publications.com

Beginners
GUIDES